NON-EUCLIDEAN GEOMETRY

A Critical and Historical Study of its Development

by ROBERTO BONOLA

Authorized English translation
with additional appendices by
H. S. CARSLAW
With an Introduction by
FEDERIGO ENRIQUES

With a supplement containing
the DR. GEORGE BRUCE HALSTED translations of
THE SCIENCE OF
ABSOLUTE SPACE
by JOHN BOLYAI

THE THEORY OF PARALLELS
by NICHOLAS LOBACHEVSKI

Dover Publications, Inc., New York

Published in Canada by General Publishing Company, Ltd., 30 Lesmill Road, Don Mills, Toronto, Ontario.

Published in the United Kingdom by Constable and Company, Ltd., 10 Orange Street, London WC 2.

This Dover edition, first published in 1955, is an unabridged and unaltered republication of the first English translation, by H. S. Carslaw, of Roberto Bonola's *La Geometria non-Euclidea* as published in 1912 by The Open Court Publishing Company. It is supplemented with George Bruce Halsted's translations of John Bolyai's Appendix to *Tentamen Juventutem Studiosam in Elementa Matheseos Purae, Elementaris ac Sublimioris, Methodo Intuitiva, Evidentia—Que Huic Propria, Introducendi* (which is here reprinted from the fourth, 1896 edition, Volume 3 of The Neomonic Series) and Nicholas Lobachevski's essay which originally appeared as *Geometrical Researches on the Theory of Parallels* in the University of Texas *Bulletin,* 1891 and is here reprinted from The Open Court Publishing Company edition of 1914.

International Standard Book Number: 0-486-60027-0
Library of Congress Catalog Card Number: 55-14932

Manufactured in the United States of America
Dover Publications, Inc.
180 Varick Street
New York, N. Y. 10014

NON-EUCLIDEAN
GEOMETRY

BY

ROBERTO BONOLA

TRANSLATED AND WITH
ADDITIONAL APPENDICES BY

H. S. CARSLAW

Introduction.

The translator of this little volume has done me the honour to ask me to write a few lines of introduction. And I do this willingly, not only that I may render homage to the memory of a friend, prematurely torn from life and from science, but also because I am convinced that the work of ROBERTO BONOLA deserves all the interest of the studious. In it, in fact, the young mathematician will find not only a clear exposition of the principles of a theory now classical, but also a critical account of the developments which led to the foundation of the theory in question.

It seems to me that this account, although concerned with a particular field only, might well serve as a model for a history of science, in respect of its accuracy and its breadth of information, and, above all, the sound philosophic spirit that permeates it. The various attempts of successive writers are all duly rated according to their relative importance, and are presented in such a way as to bring out the continuity of the progress of science, and the mode in which the human mind is led through the tangle of partial error to a broader and broader view of truth. This progress does not consist only in the acquisition of fresh knowledge, the prominent place is taken by the clearing up of ideas which it has involved; and it is remarkable with what skill the author of this treatise has elucidated the obscure concepts which have at particular periods of time presented themselves to the eyes of the investigator as obstacles, or causes of confusion. I will cite as an example his lucid analysis of the idea of there being in the case of Non-Euclidean Geometry, in contrast to Euclidean Geometry, an absolute or natural measure of geometrical magnitude.

The admirable simplicity of the author's treatment, the elementary character of the constructions he employs, the sense of harmony which dominates every part of this little work, are in accordance, not only with the artistic temperament and broad education of the author, but also with the lasting devotion which he bestowed on the Theory of Non-Euclidean Geometry from the very beginning of his scientific career. May his devotion stimulate others to pursue with ideals equally lofty the path of historical and philosophical criticism of the principles of science! Such efforts may be regarded as the most fitting introduction to the study of the high problems of philosophy in general, and subsequently of the theory of the understanding, in the most genuine and profound signification of the term, following the great tradition which was interrupted by the romantic movement of the nineteenth century.

Bologna, October 1st, 1911.

Federigo Enriques.

Translator's Preface.

BONOLA's Non-Euclidean Geometry is an elementary historical and critical study of the development of that subject. Based upon his article in ENRIQUES' collection of Monographs on Questions of Elementary Geometry[1], in its final form it still retains its elementary character, and only in the last chapter is a knowledge of more advanced mathematics required.

Recent changes in the teaching of Elementary Geometry in England and America have made it more then ever necessary that those who are engaged in the training of the teachers should be able to tell them something of the growth of that science; of the hypothesis on which it is built; more especially of that hypotheses on which rests EUCLID's theory of parallels; of the long discussion to which that theory was subjected; and of the final discovery of the logical possibility of the different Non-Euclidean Geometries.

These questions, and others associated with them, are treated in an elementary way in the pages of this book.

In the English translation, which Professor BONOLA kindly permitted me to undertake, I have introduced some changes made in the German translation.[2] For permission to do so I desire to express my sincere thanks to the firm of B. G. TEUBNER and to Professor LIEBMANN. Considerable new material has also been placed in my hands by Professor BONOLA, including a slightly altered discussion of part of

---

[1] ENRIQUES, F., *Questioni riguardanti la geometria elementare,* (Bologna, Zanichelli, 1900).

[2] Wissenschaft und Hypothese, IV. Band: *Die nichteuklidische Geometrie. Historisch-kritische Darstellung ihrer Entwicklung. Von R. Bonola. Deutsch v. H. Liebmann.* (Teubner, Leipzig, 1908).

SACCHERI's work, an Appendix on the Independence of Projective Geometry from the Parallel Postulate, and some further Non-Euclidean Parallel Constructions.

In dealing with GAUSS's contribution to Non-Euclidean Geometry I have made some changes in the original on the authority of the most recent discoveries among GAUSS's papers. A reference to THIBAUT's 'proof', and some additional footnotes have been inserted. Those for which I am responsible have been placed within square brackets. I have also added another Appendix, containing an elementary proof of the impossibility of proving the Parallel Postulate, based upon the properties of a system of circles orthogonal to a fixed circle. This method offers fewer difficulties than the others, and the discussion also establishes some of the striking theorems of the hyperbolic Geometry.

It only remains for me to thank Professor GIBSON of Glasgow for some valuable suggestions, to acknowledge the interest, which both the author and Professor LIEBMANN have taken in the progress of the translation, and to express my satisfaction that it finds a place in the same collection as HILBERT's classical *Grundlagen der Geometrie*.

P. S. As the book is passing through the press I have received the sad news of the death of Professor BONOLA. With him the Italian School of Mathematics has lost one of its most devoted workers on the Principles of Geometry. Professor ENRIQUES, his intimate friend, from whom I heard of BONOLA's death, has kindly consented to write a short introduction to the present volume. I have to thank him, and also Professor W. H. YOUNG, in whose hands, to avoid delay, I am leaving the matter of the translation of this introduction and its passage through the press.

The University, Sydney, August 1911.

H. S. Carslaw.

Author's Preface.

The material now available on the origin and development of Non-Euclidean Geometry, and the interest felt in the critical and historical exposition of the principles of the various sciences, have led me to expand the first part of my article — *Sulla teoria delle parallele e sulle geometrie non-euclidee*—which appeared six years ago in the *Questioni riguardanti la geometria elementare,* collected and arranged by Professor F. ENRIQUES.

That article, which has been completely rewritten for the German translation[1] of the work, was chiefly concerned with the systematic part of the subject. This book is devoted, on the other hand, to a fuller treatment of the history of parallels, and to the historical development of the geometries of LOBATSCHEWKY-BOLYAI and RIEMANN.

In Chapter I., which goes back to the work of EUCLID and the earliest commentators on the Fifth Postulate, I have given the most important arguments, by means of which the Greeks, the Arabs and the geometers of the Renaissance attempted to place the theory of parallels on a firmer foundation. In Chapter II., relying chiefly upon the work of SACCHERI, LAMBERT and LEGENDRE, I have tried to throw some light on the transition from the old to the new ideas, which became prevalent in the beginning of the 19th Century. In Chapters III. and IV., by the aid of the in-

---

[1] ENRIQÜES, F., Fragen der Elementargeometrie. I. Teil: Die Grundlagen der Geometrie. Deutsch von H. THIEME. (1910.) II. Teil: Die geometrischen Aufgaben, ihre Lösung und Lösbarkeit. Deutsch von H. FLEISCHER. (1907.) Teubner, Leipzig.

vestigations of GAUSS, SCHWEIKART, TAURINUS, and the con-
structive work of LOBATSCHEWSKY and BOLYAI, I have explained the principles of the first of the geometrical systems,
founded upon the denial of EUCLID's Fifth Hypothesis. In
Chapter V., I have described synthetically the further development of Non-Euclidean Geometry, due to the work of
RIEMANN and HELMHOLTZ on the structure of space, and
to CAYLEY's projective interpretation of the metrical properties of geometry.

In the whole of the book I have endeavoured to present, the various arguments in their historical order. However when such an order would have made it impossible to
treat the subject simply, I have not hesitated to sacrifice it,
so that I might preserve the strictly elementary character of
the book.

Among the numerous postulates equivalent to EUCLID's
Fifth Postulate, the most remarkable of which are brought
together at the end of Chapter IV., there is one of a *statical*
nature, whose experimental verification would furnish an
empirical foundation of the theory of parallels. In this we
have an important link between Geometry and Statics
(GENOCCHI); and as it was impossible to find a suitable place
for it in the preceding Chapters, the first of the two *Notes* [1]
in the Appendix is devoted to it.

The second *Note* refers to a theory no less interesting.
The investigations of GAUSS, LOBATSCHEWSKY and BOLYAI on
the theory of parallels depend upon an extension of one of
the fundamental conceptions of classical geometry. But a
conception can generally be extended in various directions.
In this case, the ordinary idea of parallelism, founded on
the hypothesis of non-intersecting straight lines, coplanar and

---

[1] In the English translation these *Notes* are called Appendix I.
and Appendix II.

equidistant, was extended by the above-mentioned geometers, who gave up Euclid's Fifth Postulate (equidistance),. and later, by Clifford, who abandoned the hypothesis *that the lines should be in the same plane.*

No elementary treatment of Clifford's parallels is available, as they have been studied first by the projective method (Clifford-Klein) and later, by the aid of Differential-Geometry (Bianchi-Fubini). For this reason the second Note is chiefly devoted to the exposition of their simplest and neatest properties in an elementary and synthetical manner. This Note concludes with a rapid sketch of Clifford-Klein's problem, which is allied historically to the parallelism of Clifford. In this problem an attempt is made to characterize the geometrical structure of space, by assuming as a foundation the smallest possible number of postulates, consistent with the experimental data, and with the principle of the homogeneity of space.

This is, briefly, the nature of the book. Before submitting the little work to the favourable judgment of its readers, I wish most heartily to thank my respected teacher, Professor Federigo Enriques, for the valuable advice with which he has assisted me in the disposition of the material and in the critical part of the work; Professor Corrado Segre, for kindly placing at my disposal the manuscript of a course of lectures on Non-Euclidean geometry, given by him, three years ago, in the University of Turin; and my friend, Professor Giovanni Vailati, for the valuable references which he has given me on Greek geometry, and for his help in the correction of the proofs.

Finally my grateful thanks are due to my publisher Cesare Zanichelli, who has so readily placed my book in his collection of scientific works.

Pavia, March, 1906.

Roberto Bonola.

Table of Contents.

pages

The Science of Absolute Space
and
The Theory of Parallels_____follow page 268

Chapter I.

The Attempts to prove Euclid's Parallel Postulate

The Greek Geometers and the Parallel Postulate.

§ 1. EUCLID (circa 330—275, B. C.) calls two straight lines parallel, when they are in the same plane and being produced indefinitely in both directions, do not meet one another in either direction (Def. XXIII.).[1] He proves that two straight lines are parallel, when they form with one of their transversals equal interior alternate angles, or equal corresponding angles, or interior angles on the same side which are supplementary. To prove the converse of these propositions he makes use of the following *Postulate* (V.):

If a straight line falling on two straight lines make the interior angles on the same side less than two right angles, the two straight lines, if produced indefinitely, meet on that side on which are the angles less than the two right angles.

The Euclidean Theory of Parallels is then completed by the following theorems:

Straight lines which are parallel to the same straight line are parallel to each other (Bk. I., Prop. 30).

---

[1] With regard to EUCLID's text, references are made to the critical edition of J. L. HEIBERG (Leipzig, Teubner, 1883). [The wording of this definition (XXIII), and of Postulate V below, are taken from Heath's translation of HEIBERG's text. (Camb. Univ. Press, 1908).]

Through a given point one and only one straight line can be drawn which will be parallel to a given straight line (Bk. I. Prop. 31).

The straight lines joining the extremities of two equal and parallel straight lines are equal and parallel (Bk. I. Prop. 33).

From the last theorem it can be shown that two parallel straight lines are equidistant from each other. Among the most noteworthy consequences of the Euclidean theory are the well-known theorem on the sum of the angles of a triangle, and the properties of similar figures.

§ 2. Even the earliest commentators on EUCLID's text held that Postulate V. was not sufficiently evident to be accepted without proof, and they attempted to deduce it as a consequence of other propositions. To carry out their purpose, they frequently substituted other definitions of parallels for the Euclidean definition, given *verbally* in a negative form. These alternative definitions do not appear in this form, which was believed to be a defect.

PROCLUS (410—485) — in his *Commentary on the First Book of Euclid*[1] — hands down to us valuable information upon the first attempts made in this direction. He states, for example, that POSIDONIUS (1st Century, B. C.) had proposed to call two equidistant and coplanar straight lines parallels. However, this definition and the Euclidean one correspond to two facts, which can appear separately, and

---

[1] When the text of PROCLUS is quoted, we refer to the edition of G. FRIEDLEIN: *Procli Diadochi in primum Euclidis elementorum librum commentarii*, [Leipzig, Teubner, 1873]. [Compare also W. B. FRANKLAND, *The First Book of Euclid's Elements with a Commentary based principally upon that of Proclus Diadochus*, (Camb. Univ. Press, 1905). Also HEATH's Euclid, Vol. I., Introduction, Chapter IV., to which most important work reference has been made on p. 1].

PROCLUS (p. 177), referring to a work by GEMINUS (1st Century, B. C.), brings forward in this connection the examples of the hyperbola and the conchoid, and their behaviour with respect to their asymptotes, to show that there might be parallel lines in the Euclidean sense, (that is, lines which produced indefinitely do not meet), which would not be parallel in the sense of POSIDONIUS, (that is, equidistant).

Such a fact is regarded by GEMINUS, quoting still from PROCLUS, as the most paradoxical [παραδοξότατον] in the whole of Geometry.

Before we can bring EUCLID's definition into line with that of POSIDONIUS, it is necessary to prove that if two coplanar straight lines do not meet, they are equidistant; or, that the locus of points, which are equidistant from a straight line, is a straight line. And for the proof of this proposition EUCLID requires his Parallel Postulate.

However PROCLUS (p. 364) refuses to count it among the postulates. In justification of his opinion he remarks that its converse (*The sum of two angles of a triangle is less than two right angles*), is one of the theorems proved by EUCLID (Bk. I. Prop. 17); and he thinks it impossible that a theorem whose converse can be proved, is not itself capable of proof. Also he utters a warning against mistaken appeals to self-evidence, and insists upon the (hypothetical) possibility of straight lines which are asymptotic (p. 191—2).

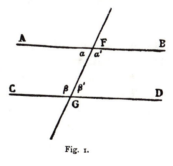

Fig. 1.

PTOLEMY (2nd Century, A. D.)—we quote again from PROCLUS (p. 362—5)—attempted to settle the question by means of the following curious piece of reasoning.

Let *AB*, *CD*, be two parallel straight lines and *FG* a transversal (Fig. 1).

Let α, β be the two interior angles to the left of *FG*, and α′, β′ the two interior angles to the right.

Then α + β will be either greater than, equal to, or less than α′ + β′.

It *is assumed* that if any one of these cases holds for one pair of parallels (e. g. α + β $>$ 2 right angles) this case will also hold for every other pair.

Now *FB*, *GD*, are parallels; as are also *FA* and *GC*.

Since α + β $>$ 2 right angles,

it follows that α′ + β′ $>$ 2 right angles.

Thus   α + β + α′ + β′ $>$ 4 right angles,

which is obviously absurd.

Hence α + β cannot be greater than 2 right angles.

In the same way it can be shown that

α + β cannot be less than 2 right angles.

Therefore we must have

α + β $=$ 2 right angles (PROCLUS, p. 365).

From this result EUCLID's Postulate can be easily obtained.

§ 3. PROCLUS (p. 371), after a criticism of Ptolemy's reasoning, attempts to reach the same goal by another path. His demonstration rests upon the following proposition, which he assumes as evident:—*The distance between two points upon two intersecting straight lines can be made as great as we please, by prolonging the two lines sufficiently.*[1]

From this he deduces the lemma: *A straight line which meets one of two parallels must also meet the other.*

---

[1] For the truth of this proposition, which he assumes as self-evident, PROCLUS relies upon the authority of ARISTOTLE. Cf. De Coelo I., 5. A rigorous demonstration of this very theorem was given by SACCHERI in the work quoted on p. 22.

His proof of this lemma is as follows:

Let *AB*, *CD*, be two parallels and *EG* a transversal, cutting the former in *F* (Fig. 2).

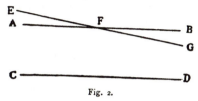

Fig. 2.

The distance of a variable point on the ray *FG* from the line *AB* increases without limit, when the distance of that point from *F* is increased indefinitely. But *since the distance between the two parallels is finite,* the straight line *EG* must necessarily meet *CD*.

PROCLUS, however, introduced the hypothesis that the distance between two parallels remains finite; and from this hypothesis EUCLID's Parallel Postulate can be logically deduced.

§ 4. Further evidence of the discussion and research among the Greeks regarding Euclid's Postulate is given by the following paradoxical argument. Relying upon it, according to PROCLUS, some held that it had been shown that two straight lines, which are cut by a third, do not meet one another, even when the sum of the interior angles on the same side is less than two right angles.

Let *AC* be a transversal of the two straight lines *AB*, *CD* and let *E* be the middle point of *AC* (Fig. 3).

On the side of *AC* on which the sum of the two internal angles is less than two right angles, take the segments *AF* and *CG* upon *AB* and *CD* each equal to *AE*. The two lines *AB* and *CD* cannot meet between the points *AF* and *CG*, since in any triangle each side is less than the sum of the other two.

The points F and G are then joined, and the same process is repeated, starting from the line FG. The segments FK and GL are now taken on AB and CD, each equal to half of FG. The two lines AB, CD are not able to meet between the points F, K and G, L.

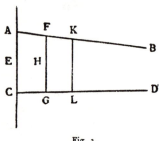

Fig. 3.

Since this operation can be repeated indefinitely, it is inferred that the two lines AB, CD will never meet.

The fallacy in this argument is contained in the use of infinity, since the segments AF, FK could tend to zero, while their sum might remain finite. The author of this paradox has made use of the principle by means of which ZENO (495—435 B. C.) maintained that it could be proved that Achilles would never overtake the tortoise, though he were to travel with double its velocity.

This is pointed out, under another form, by PROCLUS (p. 369—70), where he says that this argument proves that the point of intersection of the lines could not be reached (to determine, ὁρίζειν) by this process. It does not prove that such a point does not exist.[1]

Proclus remarks further that 'since the sum of two angles of a triangle is less than two right angles (EUCLID Bk. I. Prop. 17), there exist some lines, intersected by a third, which meet on that side on which the sum of the interior

---

[1] [Suppose we start with a triangle ABC and bisect the base BC in D. Then on BA take the segment BE equal to BD, and on CA the segment CF equal to CD, and join EF. Then repeat this process indefinitely. The vertex A can never be reached by this means, although it is at a finite distance.]

angles is less than two right angles. Thus if it is asserted that for *every* difference between this sum and two right angles the lines do not meet, it can be replied that for greater differences the lines intersect.'

'But if there exists a point of section, for *certain* pairs of lines, forming with a third interior angles on the same side whose sum is less than two right angles, it remains to be shown that this is the case for *all* the pairs of lines. *Since it might be urged that there could be a certain deficiency* (from two right angles) *for which they* (the lines) *would not intersect, while on the other hand all the other lines, for which the deficiency was greater, would intersect.*' (PROCLUS, p. 371.)

From the sequel it will appear that the question, which Proclus here suggests, can be answered in the affirmative only in the case when the segment AC of the transversal remains unaltered, while the lines rotate about the points A and C and cause the difference from two right angles to vary.

§ **5.** Another very old proof of the Fifth Postulate, reproduced in the Arabian Commentary of AL-NIRIZI[1] (9th Century), has come down to us through the Latin translation of GHERARDO DA CREMONA[2] (12th Century), and is attributed to AGANIS.[3]

The part of this commentary relating to the definitions, postulates and axioms, contains frequent references to the

---

[1] Cf. R. O. BESTHORN u. J. L. HEIBERG, 'Codex Leidensis,' 399, *1. Euclidis Elementa ex interpretatione Al-Hadschdschadsch cum commentariis Al-Narizii,* (Copenhagen, F. Hegel, 1893—97).

[2] Cf. M. CURTZE, *'Anaritii in decem libros priores elementorum Euclidis Commentarii.'* Ex interpretatione Gherardi Cremonensis in Codice Cracoviensi 569 servata, (Leipzig, Teubner, 1899).

[3] With regard to AGANIS it is right to mention that he is identified by CURTZE and HEIBERG with GEMINUS. On the other hand P. TANNERY does not accept this identification. Cf. TANNERY, *'Le philosophe Aganis est-il identique à Geminus?'* Bibliotheca Math. (3) Bd. II. p. 9—11 [1901].

the name of Sambelichius, easily identified with Simplicius, the celebrated commentator on Aristotle, who lived in the 6th Century. It would thus appear that Simplicius had written an *Introduction to the First Book of Euclid*, in which he expressed ideas similar to those of Geminus and Posidonius, affirming that the Fifth Postulate is not self evident, and bringing forward the demonstration of *his friend* AGANIS.

This demonstration is founded upon the hypothesis that equidistant straight lines exist. AGANIS calls these parallels, as had already been done by Posidonius. From this hypothesis he deduces that the shortest distance between two parallels is the common perpendicular to both the lines: that two straight lines perpendicular to a third are parallel to each other: that two parallels, cut by a third line, form interior angles on the same side, which are supplementary, and conversely.

These propositions can be proved so easily that it is unnecessary for us to reproduce the reasoning of AGANIS. Having remarked that Propositions 30 and 33 of the First Book of EUCLID follow from them, we proceed to show how AGANIS constructs the point of intersection of two straight lines which are not equidistant.

Let *AB*, *GD* be two straight lines cut by the transversal *EZ*, and such that the sum of the interior angles *AEZ*, *EZD* is less than two right angles (Fig. 4).

Without making our figure any less general we may suppose that the angle *AEZ* is a right angle.

Upon *ZD* take an arbitrary point *T*.

From *T* draw *TL* perpendicular to *ZE*.

Bisect the segment *EZ* at *P*: then bisect the segment *PZ* at *M*: and then bisect the segments *MŻ*, etc. . . . until one of the middle points *P*, *M*, . . . falls on the segment *LZ*.

Let this point, for example, be the point *M*.

Draw *MN* perpendicular to *EZ*, meeting *ZD* in *N*.

Finally from ZD cut off the segment ZC, the same multiple of ZN as ZE is of ZM.

In the case taken in the figure $ZC = 4\ ZN$.

The point C thus obtained is the point of intersection of the two straight lines AB and GD.

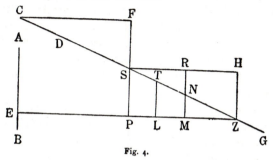

Fig. 4.

To prove this it would be necessary to show that the equal segments $ZN, NS, \ldots$, which have been cut off one after the other from the line ZD, have equal projections on ZE. We do not discuss this point, as we must return to it later (p. 11). In any case the reasoning is suggested directly by AGANIS' figure.

The distinctive feature of the preceding construction is to be noticed. It rests upon the (implicit) use of the so-called *Postulate of Archimedes*, which is *necessary* for the determination of the segment MZ, less than LZ and a submultiple of EZ.

The Arabs and the Parallel Postulate.

§ 6. The Arabs, succeeding the Greeks as leaders in mathematical discovery, like them also investigated the Fifth Postulate.

Some, however, accepted without hesitation the ideas and demonstrations of their teachers. Among this number is AL-NIRIZI (9th Century), whose commentary on the definitions, postulates and axioms of the First Book is modelled on the Introduction to the 'Elements' of SIMPLICIUS, while his demonstration of the Fifth *Euclidean Hypothesis* is that of AGANIS, to which we have above referred.

Others brought their own personal contribution to the argument. NASÎR-EDDÎN [1201—1274], for example, although in his proof of the Fifth Postulate he employs the criterion used by Aganis, deserves to be mentioned for his original idea of explicitly putting in the forefront the theorem on the sum of the angles of a triangle, and for the exhaustive nature of his reasoning.[1]

The essential part of his hypothesis is as follows: *If two straight lines r and s are the one perpendicular and the other oblique to the segment AB, the perpendiculars drawn from s upon r are less than AB on the side on which s makes an acute angle with AB, and greater on the side on which s makes an obtuse angle with AB.*

It follows immediately that if AB and $A'B'$ are two equal perpendiculars to the line BB' from the same side, the line AA' is itself perpendicular to both AB and $A'B'$. Further we have $AA' = BB'$; and therefore the figure $AA'B'B$ is a quadrilateral with its angles right angles and its opposite sides equal, i. e., a rectangle.

From this result NASÎR-EDDÎN easily deduced that the sum of the angles of a triangle is equal to two right angles. For the right-angled triangle the theorem is obvious, as it is half of a rectangle; for any triangle we obtain it by breaking up the triangle into two right-angled triangles.

With this introduction, we can now explain shortly how the Arabian geometer proves the *Euclidean Postulate* [cf. AGANIS].

---

[1] Cf.: *Euclidis elementorum libri XII studii Nassiredini*, (Rome, 1594). This work, written in Arabic, was republished in 1657 and 1801. It has not been translated into any other language.

Let *AB*, *CD* be two rays, the one oblique and the other perpendicular to the straight line *AC* (Fig. 5). From *AB* cut off the part *AH*, and from *H* draw the perpendicular *HH'* to *AC*. If the point *H'* falls on *C*, or on the opposite side of *C* from *A*, the two rays *AB* and *CD* must intersect. If, however, *H'* falls between *A* and *C*, draw the line *AL* perpendicular to *AC* and equal to *HH'*. Then, from what we have said above, *HL* = *AH'*. In *AH* produced take *HK* equal to *AH*. From *K* draw *KK'* perpendicular to *AC*. Since *KK'* > *HH'*, we can take *K'L'* = *H'H*, and join *L'H*. The

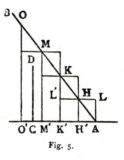

Fig. 5.

quadrilaterals *K'H'HL'*, *H'ALH* are both rectangles. Therefore the three points *L'*, *H*, *L* are in one straight line. It follows that ∢ *L'HK* = ∢ *AHL*, and that the triangles *AHL*, *HL'K* are equal. Thus *L'H* = *HL*, and from the properties of rectangles, *K'H'* = *H'A*.

In *HK* produced, take *KM* equal to *KH*. From *M* draw *MM'* perpendicular to *AC*. By reasoning similar to what has just been given, it follows that

$$M'K' = K'H' = H'A.$$

This result obtained, we take a multiple of *AH'* greater than *AC* [*The Postulate of Archimedes*]. For example, let *AO'*, equal to 4 *AH'*, be greater than *AC*. Then from *AB* cut off *AO* = 4 *AH*, and draw the perpendicular from *O* to *AC*.

This perpendicular will evidently be *OO'*. Then, in the right-angled triangle *AO'O*, the line *CD*, which is perpendicular to the side *AO'*, cannot meet the other side *OO'*, and it must therefore meet the hypotenuse *OA*.

By this means it has been proved that two straight lines *AB*, *CD*, must intersect, when one is perpendicular to the

transversal *AC* and the other oblique to it. In other words
the *Euclidean Postulate* has been proved for the çase in which
one of the internal angles is a right angle.

NASÎR-EDDÎN now makes use of the theorem on the sum
of the angles of a triangle, and by its means reduces the
general case to this particular one. We do not give his reasoning, as we shall have to describe what is equivalent to
it in a later article. [cf. p. 37.] [1]

The Parallel Postulate during the Renaissance and the 17th Century.

§ 7. The first versions of the *Elements* made in the
12th and 13th Centuries on the Arabian texts, and the later
ones, made at the end of the 15th and the beginning of the
16th, based on the Greek texts, contain hardly any critical
notes on the Fifth Postulate. Such criticism appears after the
year 1550, chiefly under the influence of the *Commentary of
Proclus*.[2] To follow this more easily we give a short sketch
of the views taken by the most noteworthy commentators of
the 16th and 17th centuries.

F. COMMANDINO [1509—1575] adds to the Euclidean
definition of parallels, without giving any justification for this

---

[1] NASÎR-EDDÎN's demonstration of the Fifth Postulate is given
in full by the English Geometer J. WALLIS, in Vol. II. of his works
(cf. Note on p. 15), and by G. CASTILLON, in a paper published in
the Mém. de l'Acad. roy. de Sciences et Belles-Lettres of Berlin,
T. XVIII. p. 175—183, (1788—1789). In addition, several other
writers refer to it, among whom we would mention chiefly, G. S.
KLÜGEL, (cf. note, (3), p. 44), J. HOFFMAN, *Kritik der Parallelentheorie*,
(Jena, 1807); V. FLAUTI, *Nuova dimostrazione del postulato quinto*, (Naples, 1818).

[2] The *Commentary of Proclus* was first printed at Basle (1533)
in the original text; and next at Padua (1560) in Barozzi's Latin
translation.

step, the idea of equidistance. With regard to the Fifth Postulate he gives the views and the demonstration of PROCLUS.[1]

C. S. CLAVIO [1537—1612], in his Latin translation of Euclid's text[2], reproduces and criticises the demonstration of PROCLUS. Then he brings forward a new demonstration of the Euclidean hypothesis, based on the theorem: *The line equidistant from a straight line is a straight line*; which he attempts to justify by similar reasoning. His demonstration has many points in common with that of Nasîr-Eddîn.

P. A. CATALDI [?—1626] is the first modern mathematician to publish a work devoted exclusively to the theory of parallels.[3] CATALDI starts from the conception of equidistant and non-equidistant straight lines; but to prove the effective existence of equidistant straight lines, he adopts the hypothesis that *straight lines which are not equidistant converge in one direction and diverge in the other.* [cf. NASÎR-EDDÎN.][4].

G. A. BORELLI [1608—1679] takes the following Axiom [XIV], and attempts to justify his assumption:

'*If a straight line which remains always in the same plane as a second straight line, moves so that the one end always touches this line, and during the whole displacement the first remains continually perpendicular to the second, then the other end, as it moves, will describe a straight line.*'

Then he shows that two straight lines which are perpendicular to a third are equidistant, and he defines parallels as equidistant straight lines.

The theory of parallels follows.[5]

---

[1] *Elementorum libri XV,* (Pesaro, 1572).

[2] *Euclidis elementorum libri XV,* (Rome, 1574).

[3] *Operetta delle linee rette equidistanti et non equidistanti,* (Bologna, 1603).

[4] CATALDI made some further additions to his argument in the work, *Aggiunta all' operetta delle linee rette equidistanti et non equidistanti.* (Bologna, 1604).

[5] BORELLI: *Euclides restitutus,* (Pisa, 1658).

§ 8. Giordano Vitale [1633 — 1711] again returns to
the idea of equidistance put forward by Posidonius, and re-
cognizes, with Proclus, that it is necessary to exclude the possibility of the Euclidean parallels being asymptotic lines. To
this end he defines two equidistant straight lines as parallels,
and attempts to prove that the locus of the points equidistant
from one straight line is another straight line.[1]

His demonstration practically depends upon the following lemma:

*If two points, A, C upon a curve, whose concavity is to-
wards X, are joined by the straight line AC, and perpendiculars
are drawn from the infinite number of points of the arc AC
upon any straight line, then these perpendiculars cannot be equal
to each other.*

The words 'any straight line', in this enunciation, do not
refer to a straight line taken at random in the plane, but to

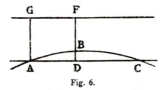
Fig. 6.

a straight line constructed in
the following way (Fig. 6).
From the point *B* of the arc
AC draw *BD* perpendicular to
the chord *AC.* Then at *A* draw
AG also perpendicular to *AC.*

Finally, having cut off equal segments *AG* and *DF* upon
these two perpendiculars, join the ends *G* and *F. GF* is the
straight line which Giordano considers in his demonstration,
a straight line with respect to which the arc *AB* is certainly
not an equidistant line.

But when the author wishes to prove that the locus of
points equidistant from a straight line is also a straight line,
he applies the preceding lemma to a figure in which the relations existing between the arc *ABC* and the straight line

---

[1] Giordano Vitale: *Euclide restituto overo gli antichi elementi
geometrici ristaurati, e facilitati. Libri XV.* (Rome, 1680).

GF do not hold. Thus the consequences which he deduces from the existence of equidistant straight lines are not really legitimate.

From this point of view GIORDANO's proof makes no advance upon those which preceded it. However it includes a most remarkable theorem, containing an idea which will be further developed in the articles which follow.

Let $ABCD$ be a quadrilateral of which the angles A, B are right angles and the sides AD, BC equal (Fig. 7). Further, let HK be the perpendicular drawn from a point H, upon the side DC, to the base AB of the quadrilateral. GIORDANO proves: (i) that the angles D, C are equal; (ii) that, when the segment HK is equal to the segment AD, the

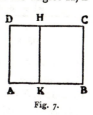
Fig. 7.

two angles D, C are right angles, and CD is equidistant from AB.

By means of this theorem GIORDANO reduces the question of equidistant straight lines to the proof of the existence of *one point H* upon DC, whose distance from AB is equal to the segments AD and BC. We regard this as one of the most noteworthy results in the theory of parallels obtained up to that date.[1]

§ 9. J. WALLIS [1616—1703] abandoned the idea of equidistance, employed without success by the preceding mathematicians, and gave a new demonstration of the Fifth Postulate. He based his proof on the Axiom: *To every figure there exists a similar figure of arbitrary magnitude.* We now describe shortly how WALLIS proceeds:[2]

---

[1] Cf.: BONOLA: *Un teorema di Giordano Vitale da Bitonto sulle rette equidistanti*, Bollettino di Bibliografia e Storia delle Scienze Mat. (1905).

[2] Cf.: WALLIS: *De Postulato Quinto; et Definizione Quinta; Lib. 6.*

Let a, b be two straight lines intersected at A, B by the transversal c (Fig. 8). Let α, β be the interior angles on the

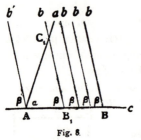

Fig. 8.

same side of c, such that α + β is less than two right angles. Through A draw the straight line b' so that b and b' form with c equal corresponding angles. It is clear that b' will lie in the angle adjacent to α. Let the line b be now moved continuously along the segment AB, so that the angle which it makes with c remains always equal to β. Before it reaches its final position b' it must necessarily intersect a. In this way a triangle $AB_1 C_1$ is determined, with the angles at A and B_1 respectively equal to α and β.

But, by WALLIS's hypothesis of the existence of similar figures, upon AB, the side homologous to AB_1, we must be able to construct a triangle ABC similar to the triangle $AB_1 C_1$. This is equivalent to saying that the straight lines a, b must meet in a point, namely, the third angular point of the triangle ABC. Therefore, etc.

Wallis then seeks to justify the new position he has taken up. He points out that Euclid, in postulating the existence of a circle of given centre and given radius, [Post. III.], practically admits the principle of similarity for circles. But even although intuition would support this view, the idea of form, independent of the dimensions of the figure, constitutes a

Euclidis; disceptatio geometrica. Opera Math. t. II, p. 669—78 (Oxford, 1693). This work by WALLIS contains two lectures given by him in the University of Oxford; the first in 1651, the second in 1663. It also contains the demonstration of NASÎR-EDDÎN. The part containing WALLIS's proof was translated into German by ENGEL and STÄCKEL in their *Theorie der Parallellinien von Euclid bis auf Gauss*, p. 21—36, (Leipzig, Teubner, 1895). We shall quote this work in future as *Th. der P.*

hypothesis, which is certainly not more evident than the Postulate of EUCLID.

We remark, further, that WALLIS could more simply have assumed the existence of triangles with equal angles, or, as we shall see below, of only two unequal triangles whose angles are correspondingly equal.

[cf. p. 29 Note 1.]

§ 10. The critical work of the preceding geometers is sufficient to show the historical development of our subject in the 16th and 17th Centuries, so that it would be superfluous to speak of other able writers, such as, e. g., OLIVER of BURY [1604], LUCA VALERIO [1613], H. SAVILE [1621], A. TACQUET [1654], A. ARNAULD [1667].[1] However, it seems necessary to say a few words on the question of the position which the different commentators on the 'Elements' allot to the Euclidean hypothesis in the system of geometry.

In the Latin edition of the 'Elements' [1482], based upon the Arabian texts, by CAMPANUS [13th Century], this hypothesis finds a place among the postulates. The same may be said of the Latin translation of the Greek version by B. ZAMBERTI [1505], of the editions of LUCA PACIUOLO [1509], of N. TARTAGLIA [1543], of F. COMMANDINO [1572], and of G. A. BORELLI [1658].

On the other hand the first printed copy of the 'Elements' in Greek, [Basle, 1533], contains the hypothesis among the axioms [Axiom XI]. In succession it is placed among the Axioms by F. CANDALLA [1556], C. S. CLAVIO [1574], GIORDANO VITALE [1680], and also by GREGORY [1703], in his well-known Latin version of EUCLID's works.

To attempt to form a correct judgment upon these dis-

---

[1] For fuller information on this subject cf. RICCARDI: *Saggio di una bibliografia euclidea*. Mem. di Bologna, (5) T. I. p. 27—34, (1890).

crepancies, due more to the manuscripts handed down from the Greeks than to the aforesaid authors, it will be an advantage to know what meaning the former gave to the words *'postulates'* [αἰτήματα] and *'axioms'* [ἀξιώματα].[1] First of all we note that the word *'axioms'* is used here to denote what Euclid in his text calls *'common notions'* [κοιναὶ ἔννοιαι].

PROCLUS gives three different ways of explaining the difference between the axioms and postulates.

The first method takes us back to the difference between a *problem* and a *theorem*. A *postulate* differs from an *axiom*, as a *problem* differs from a *theorem*, says PROCLUS. By this we must understand that *a postulate affirms the possibility of a construction.*

The second method consists in saying that *a postulate is a proposition with a geometrical meaning, while an axiom is a proposition common both to geometry and to arithmetic.*

Finally the third method of explaining the difference between the two words, given by PROCLUS, is supported by the authority of ARISTOTLE [384—322 B. C.]. The words *axiom* and *postulate* do not appear to be used by ARISTOTLE exclusively in the mathematical sense. *An axiom is that which is true in itself,* that is, owing to the meaning of the words which it contains; *a postulate is that which, although it is not an axiom, in the aforesaid sense, is admitted without demonstration.*

Thus the word axiom, as is more evident from an example due to ARISTOTLE, [*when equal things are subtracted from equal things the remainders are equal*], is used in a sense which

---

[1] For the following, cf. PROCLUS, in the chapter entitled *Petita et axiomata.* In a Paper read at the Third Mathematical Congress (Heidelberg, 1904) G. VAILATI has called the attention of students anew to the meaning of these words among the Greeks. Cf.: *Intorno al significato della distinzione tra gli assiomi ed i postulati nella geometria greca.* Verh. des dritten Math. Kongresses, p. 575—581, (Leipzig, Teubner, 1905).

corresponds, at any rate very closely, to that of the *common notions* of EUCLID, whilst the word *postulate* in ARISTOTLE has a different meaning from each of the two to which reference has just been made.[1]

Hence according as one or other of these distinctions between the words is adopted, a particular proposition would be placed among the *postulates* or among the *axioms*. If we adopt the first, only the first three of the five postulates of EUCLID, according to PROCLUS, have a right to this name, since only in these are we asked to carry out a construction [to join two points, to produce a straight line, to describe a circle whose centre and radius are arbitrary]. On the other hand, Postulate IV. [all right angles are equal], and Postulate V. ought to be placed among the axioms.[2]

---

[1] Cf. ARISTOTLE: *Analytica Posteriora.* I. 10. § 8. We quote in full this slightly obscure passage, where the philosopher speaks of the postulate: ὅσα μὲν οὖν δεικτὰ ὄντα λαμβάνει αὐτὸς μὴ δείξας, ταῦτα ἐὰν μὲν δοκοῦντα λαμβάνῃ τῷ μανθάνοντι ὑποτίθεται. Καὶ ἔστιν οὐχ ἁπλῶς ὑπόθεσις ἀλλὰ πρὸς ἐκεῖνον μόνον. Ἐὰν δὲ ἢ μηδεμιᾶς ἐνούσης δόξης ἢ καὶ ἐναντίας ἐνούσης λαμβάνῃ, τὸ αὐτὸ αἰτεῖται. Καὶ τούτῳ διαφέρει ὑπόθεσις καὶ αἴτημα, ἔστι γὰρ αἴτημα τὸ ὑπεναντίον τοῦ μανθάνοντος τῇ δόξῃ.

[2] It is right to remark that the Fifth Postulate can be enunciated thus: *The common point of two straight lines can be found, when these two lines, cut by a transversal, form two interior angles on the same side whose sum is less than two right angles.* Thus it follows that this postulate affirms, like the first three, the possibility of a construction. However this character disappears altogether, if it is enunciated, for example, thus: *Through a point there passes only one parallel to a straight line;* or, thus: *Two straight lines which are parallel to a third line are parallel to each other.* It would therefore appear that the distinction noted above is purely formal. However we must not let ourselves be deceived by appearances. The Fifth Postulate, in whatever way it is enunciated, practically allows the construction of the point of intersection of all the straight lines of a pencil with a given straight line in the plane of the pencil, one of these lines alone being excepted. It is true that there is a certain

Again, if we accept the second or the third distinction, the five Euclidean postulates should all be included among the postulates.

In this way the origin of the divergence between the various manuscripts is easily explained. To give greater weight to this explanation we might add the uncertainty which historians feel in attributing to EUCLID the *postulates, common notions* and *definitions* of the First Book. So far as regards the postulates, the gravest doubts are directed against the last two. The presence of the first three is sufficiently in accord with the whole plan of the work.[1] Admitting the hypothesis that the Fourth and Fifth Postulates are not Euclid's, even if it is against the authority of Geminus and Proclus, the extreme rigour of the '*Elements*' would naturally lead the later geometers to seek in the body of the work all those propositions which are admitted without demonstration. Now the one which concerns us is found stated very concisely in the demonstration of Bk. I. Prop. 29. From this, the substance of the Fifth Postulate could then be taken, and added to the postulates of construction, or to the axioms, according to the views held by the transcriber of EUCLID's work.

Further, its natural place would be, and this is GREGORY's view, after Prop. 27, of which it enunciates the converse.

Finally, we remark that, whatever be the manner of deciding the verbal question here raised, the modern philosophy of mathematics is inclined generally to suppress the

---

difference between this postulate and the three postulates of construction. In the latter the data are completely independent. In the former the data (the two straight lines cut by a transversal) are subject to a condition. So that the Euclidean Hypothesis belongs to a class intermediate between the postulates and axiom, rather than to the one or the other.

[1] Cf. P. TANNERY: *Sur l'authenticité des axiomes d'Euclide.* Bull. d. Sc. Math. (2), T. VIII. p. 162—175, (1884).

distinction between postulate and axiom, which is adopted in the second and third of the above methods. The generally accepted view is to regard the fundamental propositions of geometry as hypotheses resting upon an empirical basis, while it is considered superfluous to place statements, which are simple consequences of the given definitions, among the propositions.

Chapter II.

The Forerunners of Non-Euclidean Geometry.

Gerolamo Saccheri [1667—1733].

§ 11. The greater part of the work of GEROLAMO SAC-CHERI: *Euclides ab omni naevo vindicatus: sive conatus geo-metricus quo stabiliuntur prima ipsa universae Geometriae Principia*, [Milan, 1733], is devoted to the proof of the Fifth Postulate. The distinctive feature of SACCHERI's geometrical writings is to be found in his '*Logica demonstrativa*', [Turin, 1697]. It is simply a particular method of reasoning, already used by EUCLID [Bk. IX. Prop. 12], according to which *by assuming as hypothesis that the proposition which is to be proved is false, one is brought to the conclusion that it is true.*[1]

Adopting this idea, the author takes as data the first twenty-six propositions of EUCLID, and he assumes as a hypothesis that the Fifth Postulate is false. Among the consequences of this hypothesis he seeks for some proposition, which would entitle him to affirm the truth of the postulate itself.

Before entering upon an exposition of SACCHERI's work, we note that EUCLID assumes implicitly that the straight line is *infinite* in the demonstration of Bk. I. 16 [the exterior angle of a triangle is greater than either of the interior and opposite

---

[1] Cf. G. VAILATI: *Di un' opera dimenticata del P. Gerolamo Sac-cheri*, Rivista Filosofica (1903).

angles], since his argument is practically based upon the existence of a segment which is double a given segment.

We shall deal later with the possibility of abandoning this hypothesis. At present we note that SACCHERI tacitly assumes it, since in the course of his work he uses the *proposition of the exterior angle.*

Finally, we note that he also employs the *Postulate of Archimedes* [1] and the *hypothesis of the continuity of the straight line,* [2] to extend, to all the figures of a given type, certain propositions admitted to be true only for a single figure of that type.

§ 12. The fundamental figure of SACCHERI is the *two right-angled isosceles* quadrilateral; that is, the quadrilateral of which two opposite sides are equal to each other and perpendicular to the base. The properties of such a figure are deduced from the following Lemma I., which can easily be proved:

If a quadrilateral ABCD has the consecutive angles A and B right angles, and the sides AD and BC equal, then the angle C is equal to the angle D [This is a special case of SACCHERI's Prop. I.]; *but if the sides AD and BC are unequal, of the two angles C, D, that one is greater which is adjacent to the shorter side, and vice versa.*

---

[1] [The Postulate of Archimedes is stated by Hilbert thus: Let A_1 be any point upon a straight line between the arbitrarily chosen points A and B. Take the points A_2, A_3, . . . so that A_1 lies between A and A_2, A_2 between A_1 and A_3, etc.; moreover let the segments AA_1, A_1A_2, A_2A_3, . . . be all equal. Then among this series of points, there always exists a certain point A_n, such that B lies between A and A_n.]

[2] This hypothesis is used by SACCHERI in its intuitive form, viz.: a segment, which passes continuously from the length a to the length b, different from a, takes, during its variation, every length intermediate between a and b.

Let *ABCD* be a quadrilateral with two right angles *A* and *B*, and two equal sides *AD* and *BC* (Fig. 9). On the *Euclidean hypothesis* the angles *C* and *D* are also right angles. Thus, if we assume that they are able to be both *obtuse*, or both *acute*, we implicitly deny the Fifth Postulate. SACCHERI discusses these three hypotheses regarding the angles *C, D*. He named them:

The Hypothesis of the Right Angle
$$[\measuredangle C = \measuredangle D = \text{1 right angle}]:$$
The Hypothesis of the Obtuse Angle
$$[\measuredangle C = \measuredangle D > \text{1 right angle}]:$$
The Hypothesis of the Acute Angle
$$[\measuredangle C = \measuredangle D < \text{1 right angle}].$$

One of his first important results is the following:

According as the Hypothesis of the Right Angle, of the Obtuse Angle, or of the Acute Angle is true in the two right-angled isosceles quadrilateral, we must have AB = CD, AB > CD, or AB < CD, respectively. [Prop. III.]

In fact, on the *Hypothesis of the Right Angle*, by the preceding Lemma, we have immediately

$$AB = CD.$$

On the *Hypothesis of the Obtuse Angle*, the perpendicular

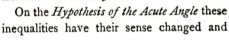

OO' at the middle point of the segment *AB* divides the fundamental quadrilateral into two equal quadrilaterals, with right angles at O and O'. Since the angle $D >$ angle A, then we must have $AO > DO'$, by this Lemma. Thus $AB > CD$.

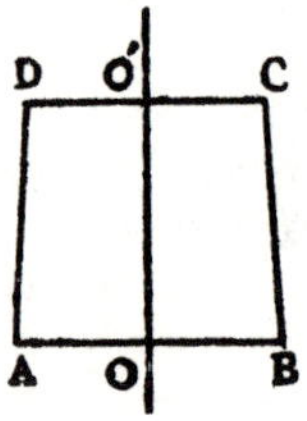

On the *Hypothesis of the Acute Angle* these inequalities have their sense changed and we have

Fig 9.

$$AB < CD.$$

Using the reductio ad absurdum argument, we obtain the converse of this theorem. [Prop. IV.]

If the Hypothesis of the Right Angle is true in only one case, then it is true in every other case. [Prop. V.]

Suppose that in the two right-angled isosceles quadrilateral *ABCD* the *Hypothesis of the Right Angle* is verified.

In *AD* and *BC* (Fig. 10) take the points *H* and *K* equidistant from *AB*; join *HK* and form the quadrilateral *ABKH*.

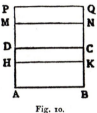

Fig. 10.

If *HK* is perpendicular to *AH* and *BK*, the *Hypothesis of the Right Angle* is also verified in the new quadrilateral.

If it is not, suppose that the angle *AHK* is acute. Then the adjacent angle *DHK* is obtuse. Thus in the quadrilateral *ABKH*, from the *Hypothesis of the Acute Angle*, it follows that $AB < HK$: while in the quadrilateral *HKCD*, from the *Hypothesis of the Obtuse Angle*, it follows that $HK < CD$.

But these two inequalities are contradictory, since by the *Hypothesis of the Right Angle* in the quadrilateral *ABCD*, $AB = CD$.

Thus the angle *AHK* cannot be acute: and since by the same reasoning we could prove that the angle *AHK* cannot be obtuse, it follows that the *Hypothesis of the Right Angle* is also true in the quadrilateral *ABKH*.

On *AD* and *BC* produced, take the points *M, N* equidistant from the base *AB*. Then the *Hypothesis of the Right Angle* is also true for the quadrilateral *ABNM*. In fact if *AM* is a multiple of *AD*, the proposition is obvious. If *AM* is not a multiple of *AD*, we take a multiple of *AD* greater than *AM* [*the Postulate of Archimedes*], and from *AD* and *BC* produced cut off *AP* and *BQ* equal to this multiple. Since, as we have just seen, the *Hypothesis of the Right Angle* is true in the quadrilateral *ABQP*, the same hypothesis must also hold in the quadrilateral *ABNM*.

Finally the said hypothesis must hold for a quadrilateral

on any base, since, in Fig. 10, we can take as the base one
of the sides perpendicular to AB.

Note. This theorem of SACCHERI is practically contained
in that of GIORDANO VITALE, stated on p. 15. In fact, referring to Fig. 7, the hypothesis

$$DA = HK = CB$$

is equivalent to the other

$$\angle D = \angle H = \angle C = 1 \text{ right angle.}$$

But from the former, there follows the equidistance of the
two straight lines DC, AB[1]; and thus the validity of the *Hypothesis of the Right Angle* in all the two right-angled isosceles
quadrilaterals, whose altitude is equal to the line DA, is
established. The same hypothesis is also true in a quadrilateral of any height, since the line called at one time the
base may later be regarded as the height.

*If the Hypothesis of the Obtuse Angle is true in only one
case, then it is true in every other case.* [Prop. VI.]

Referring to the standard quadrilateral $ABCD$ (Fig. 11),

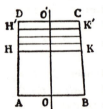

Fig. 11.

suppose that the angles C and D are obtuse. Upon AD and BC take the points
H and K equidistant from AB.

In the first place we note that the
segment HK cannot be perpendicular to
the two sides AD and BC, since in that
case the *Hypothesis of the Right Angle*
would be verified in the quadrilateral
$ABKH$, and consequently in the fundamental quadrilateral.

Let us suppose that the angle AHK is acute. Then

---

[1] It is true that GIORDANO in his argument refers to the points
of the segment DC, which he shows are equidistant from the base
AB of the quadrilateral. However the same argument is applicable
to all the points which lie upon DC, or upon DC produced. Cf.
BONOLA's Note referred to on p. 15.

by *the Hypothesis of the Acute Angle, HK > AB*. But as the
Hypothesis of the Obtuse Angle holds in *ABDC,* we have
$$AB > CD.$$
Therefore $HK > AB > CD.$
If we now move the straight line *HK continuously,* so that it
remains perpendicular to the median *OO'* of the fundamental
quadrilateral, the segment *HK,* contained between the opposite sides *AD, BC,* which in its initial position is greater than
AB, will become less than *AB* in its final position *DC.* From
the postulate of continuity we may then conclude that,
between the initial position *HK* and the final position *DC,*
there must exist an intermediate position *H'K',* for which
$H'K' = AB.$

Consequently in the quadrilateral *ABK'H'* the *Hypothesis of the Right Angle* would hold [Prop. III.], and therefore,
by the preceding theorem, the *Hypothesis of the Obtuse Angle*
could not be true in *ABCD.*

The argument is also valid if the segments *AH, BK* are
greater than *AD,* since it is impossible that the angle *AHK*
could be acute. Thus the *Hypothesis of the Obtuse Angle* holds
in *ABKH* as well as in *ABCD.*

Having proved the theorem for a quadrilateral whose
sides are of any size, we proceed to prove it for one whose
base is of any size: for example the base *BK* [cf. Fig. 12].

Since the angles *K, H,* are obtuse, the
perpendicular at *K* to *KB* will meet the
segment *AH* in the point *M,* making the
angle *AMK* obtuse [theorem of the exterior angle].

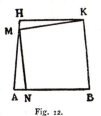

Fig. 12.

Then in *ABKM* we have $AB > KM$,
by Lemma I. Cut off from *AB* the segment
BN equal to *MK.* Then we can construct
the two right-angled isosceles quadrilateral *BKMN,* with the
angle *MNB* obtuse, since it is an exterior angle of the triangle

ANM. It follows that the *Hypothesis of the Obtuse Angle* holds in the new quadrilateral.

Thus the theorem is completely demonstrated.

If the Hypothesis of the Acute Angle is true in only one case, then it is true in every other case. [Prop. VII.]

This theorem can be easily proved by using the method of reductio ad absurdum.

§ 13. From the theorems of the last article SACCHERI easily obtains the following important result with regard to triangles:

According as the Hypothesis of the Right Angle, the Hypothesis of the Obtuse Angle, or the Hypothesis of the Acute Angle, is found to be true, the sum of the angles of a triangle will be respectively equal to, greater than, or less than two right angles. [Prop. IX.]

Let *ABC* [Fig. 13] be a triangle of which *B* is a right angle. Complete the quadrilateral by drawing *AD* perpendicular to *AB* and equal to *BC*; and jon *CD*.

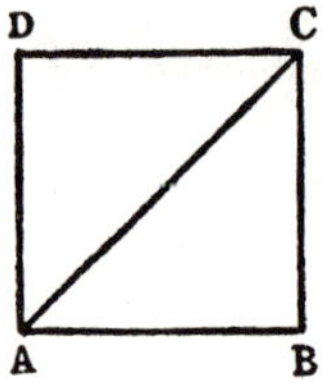

Fig. 13.

On the Hypothesis of the Right Angle, the two triangles *ABC* and *ADC* are equal. Therefore $\angle BAC = \angle DCA$.

It follows immediately that in the triangle *ABC*,

$$\angle A + \angle B + \angle C = 2 \text{ right angles.}$$

On the Hypothesis of the Obtuse Angle,
since $AB > DC$,
we have $\angle ACB > \angle DAC$.[1]

[1] This inequality is proved by SACCHERI in his Prop. VIII., and serves as Lemma to Prop. IX. It is, of course, Prop. 25 of EUCLID's First Book.

Therefore, in this triangle we shall have

$$\angle A + \angle B + \angle C > 2 \text{ right angles.}$$

On the Hypothesis of the Acute Angle,
since $AB < DC$,
we have $\angle ACB < \angle DAC$,

and therefore, in the same triangle,

$$\angle A + \angle B + \angle C < 2 \text{ right angles.}$$

The theorem just proved can be easily extended to the case of any triangle, by breaking the figure up into two right angled triangles. In Prop. XV. SACCHERI proves the converse, by a reductio ad absurdum.

The following theorem is a simple deduction from these results:

If the sum of the angles of a triangle is equal to, greater than, or less than two right angles in only one triangle, this sum will be respectively equal to, greater than, or less than two right angles in every other triangle.[1]

This theorem, which SACCHERI does not enunciate explicitly, Legendre discovered anew and published, for the first and third hypotheses, about a century later.

§ 14. The preceding theorems on the two right-angled isosceles quadrilaterals were proved by SACCHERI, and

---

[1] Another of SACCHERI's propositions, which does not concern us directly, states that *if the sum of the angles of only one quadrilateral is equal to, greater than, or less than four right angles, the Hypothesis of the Right Angle, the Hypothesis of the Obtuse Angle, or the Hypothesis of the Acute Angle would respectively be true.* A note of SACCHERI's on the Postulate of WALLIS (cf. § 9) makes use of this proposition. He points out that WALLIS needed only to assume the existence of two triangles, whose angles were equal each to each and sides unequal, to deduce the existence of a quadrilateral in which the sum of the angles is equal to four right angles. From this the validity of the *Hypothesis of the Right Angle* would follow, and in its turn the Fifth Postulate.

later by other geometers, with the help of the *Postulate of
Archimedes* and the *principle of continuity* [cf. Prop. V., VI].
However DEHN[1] has shown that they are independent of
these hypotheses. This can also be proved in an elementary
way as follows.[2]

On the straight line r (Fig. 14) let two points B and D
be chosen, and equal perpendiculars BA and DC be drawn
to these lines. Let A and C be joined by the straight line s.
The figure so obtained, in which evidently $\sphericalangle BAC = \sphericalangle DCA$,
is fundamental in our argument and we shall refer to it con-
stantly.

Two points E, E' are now taken on s, of which the
first is situated between A and C, and the second not.

Further let the perpendiculars from E, E' to the line
r meet it at F and F'.

The following theorems now hold:

I. $\begin{cases} \text{If } EF = AB, \\ \quad\text{or} \\ E'F' = AB \end{cases}$, the angles BAC, DCA are right angles.

II. $\begin{cases} \text{If } EF > AB, \\ \quad\text{or} \\ E'F' < AB \end{cases}$, the angles BAC, DCA are obtuse.

III. $\begin{cases} \text{If } EF < AB, \\ \quad\text{or} \\ E'F' > AB \end{cases}$, the angles BAC, DCA are acute.

We now prove Theorem I. [cf. Fig. 14.]

From the hypothesis $EF = AB$, the following equalities
are deduced:

---

[1] Cf. *Die Legendreschen Sätze über die Winkelsumme im Dreieck.*
Math. Ann. Bd. 53, p. 405—439 (1900).

[2] Cf. BONOLA, *1 teoremi del Padre Gerolamo Saccheri sulla
somma degli angoli di un triangolo e le ricerche di M. Dehn.* Rend.
Istituto Lombardo (2), Vol. XXXVIII. (1905).

$$\angle BAE = \angle FEA, \text{ and } \angle FEC = \angle DCE.$$

These, together with the fundamental equality

$$\angle BAC = \angle DCA,$$

are sufficient to establish the equality of the two angles FEA and FEC.

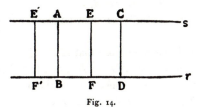

Fig. 14.

Since these are adjacent angles, they are both right angles, and consequently the angles BAC and DCA are right angles.

The same argument is applicable in the hypothesis

$$E'F' = AB.$$

We proceed to Theorem II [cf. Fig. 15].

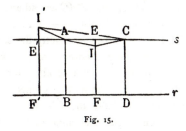

Fig. 15.

Suppose, in the first place, $EF > AB$. From FE cut off $FI = AB$, and join I to A and C.

Then the following equalities hold:

$$\angle BAI = \angle FIA \text{ and } \angle DCI = \angle FIC.$$

Further, by the theorem of the exterior angle [Bk. I. 16], we have

$$\angle FIA + \angle FIC > \angle FEA + \angle FEC = 2 \text{ right angles.}$$

But

$$\angle BAC + \angle DCA > \angle BAI + \angle DCI.$$

Therefore

$$\angle BAC + \angle DCA > \angle FIA + \angle FIC > 2 \text{ right angles.}$$

But, since $\angle BAC = \angle DCA$,

it follows that $\angle BAC > 1$ right angle. . . . Q. E. D.

In the second place, suppose that $E'F' < AB$. Then from $F'E'$ produced cut off $F'I' = BA$, and join I' to C and A.

The following relations, as usual, hold:

$$\angle F'I'A = \angle BAI', \ \angle F'I'C' = \angle DCI';$$
$$\angle I'AE' > \angle I'CE', \ \angle F'I'A < \angle F'I'C.$$

Combining these results, we deduce, first of all, that

$$\angle BAI' < \angle DCI'.$$

From this, if we subtract the terms of the inequality

$$\angle I'AE' > \angle I'CE',$$

we obtain

$$\angle BAE' < \angle DCE' = \angle BAC.$$

But the two angles BAE' and BAC are adjacent. Thus we have proved that $\angle BAC$ is obtuse.—Q. E. D.

Theorem III. can be proved in exactly the same way.

The converses of these theorems can now be easily shown to be true by the reductio ad absurdum method. In particular, if M and N are the middle points ot the two segments AC and BD, we have the following results for the segment MN which is perpendicular to both the lines AC and BD (Fig. 16).

If $\angle BAC = \angle DCA = 1$ *right angle, then* $MN = AB$.
If $\angle BAC = \angle DCA > 1$ *right angle, then* $MN > AB$.
If $\angle BAC = \angle DCA < 1$ *right angle, then* $MN < AB$.

Further it is easy to see that

(i)   *If* $\angle BAC = \angle DCA = 1$ *right angle,*
then $\angle FEM$ *and* $\angle F'E'M$ *are each 1 right angle.*

(ii) *If $\angle BAC = \angle DCA >$ I right angle,
then $\angle FEM$ and $\angle F'E'M$ are each obtuse.*

(iii) *If $\angle BAC = \angle DCA <$ I right angle,
then $\angle FEM$ and $\angle F'E'M$ are each acute.*

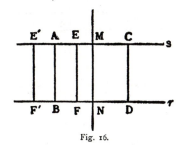

Fig. 16.

In fact, in Case (i), since the lines *r* and *s* are equidistant, the following equalities hold:

$\angle NMA = \angle FEM = \angle BAC = \angle F'E'M =$ I right angle.

To prove Cases (ii) and (iii), it is sufficient to use the reductio ad absurdum method, and to take account of the results obtained above.

Now let *P* be a point on the line *MN*, not contained between *M* and *N* (Fig. 17). Let *RP* be the perpendicular to *MN* and *RK* the perpendicular to *BD*. This last perpendicular will meet *AC* in a point *H*. On this understanding the preceding theorems immediately establish the truth of the following results:

*If $\angle BAM =$ I right angle, then $\angle KHM$ and $\angle KRP$
are each equal to I right angle.*

*If $\angle BAM >$ I right angle, then $\angle KHM$ and $\angle KRP$
are each greater than I right angle.*

*If $\angle BAM <$ I right angle, then $\angle KHM$ and $\angle KRP$
are each less than I right angle.*

These results are also true, as can easily be seen, if the point *P* falls between *M* and *N*.

In conclusion, the last three theorems, which clearly

coincide with Saccheri's theorems upon the two right-angled isosceles quadrilateral, are equivalent to the following result, proved without using Archimedes' Postulate:—

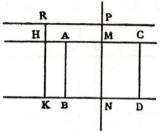

Fig. 17.

If the truth of the Hypothesis of the Right Angle, of the Obtuse Angle, or of the Acute Angle, respectively, is known in only one case, its truth is also known in every other case.

If we wish now to pass from the theorems on quadrilaterals to the corresponding theorems on triangles, we need only refer to SACCHERI's demonstration [cf. p. 28], since this part of his argument does not in any way depend upon the postulate in question.

We have thus obtained the result which was to be proved.

§ 15. To make our exposition of SACCHERI's work more concise, we take from Prop. XI. and XII. the contents of the following Lemma II:

Let ABC be a triangle of which C is a right angle: let H be the middle point of AB, and K the foot of the perpendicular from H upon AC. Then we shall have

AK = KC, on the Hypothesis of the Right Angle;

AK < KC, on the Hypothesis of the Obtuse Angle;

AK > KC, on the Hypothesis of the Acute Angle.

On *the Hypothesis of the Right Angle* the result is obvious.

On the Hypothesis of the Obtuse Angle, since the sum of the angles of a quadrilateral is greater than four right angles, it follows that $\sphericalangle AHK < \sphericalangle HBC$. Let HL be the perpendicular from H to BC (Fig. 18). Then the result just obtained, and the fact that the two triangles AHK, HBL have equal hypotenuses, give rise to the following inequality: $AK < HL$. But the quadrilateral $HKCL$ has three right angles and therefore the angle H is obtuse [*Hypothesis of the Obtuse Angle*]. It follows that

$$HL < KC,$$

and thus

$$AK < KC.$$

The third part of this Lemma can be proved in the same way.

It is easy to extend this Lemma as follows (Fig. 19):

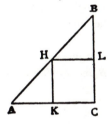

Fig. 18.

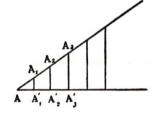

Fig. 19.

Lemma III. If on the one arm of an angle A equal segments AA_1, $A_1 A_2$, $A_2 A_3$, . . . are taken, and AA_1', $A_1' A_2'$, $A_2' A_3'$ . . . are their projections upon the other arm of the angle, then the following results are true:

$$AA_1' = A_1' A_2' = A_2' A_3' = . . .$$

on the Hypothesis of the Right Angle;

$$AA_1' < A_1' A_2' < A_2' A_3' = < . . .$$

on the Hypothesis of the Obtuse Angle;

$$AA_1' > A_1' A_2' > A_2' A_3' > . . .$$

on the Hypothesis of the Acute Angle.

To save space the simple demonstration is omitted.

We can now proceed to the proof of Prop. XI. and XII. of Saccheri's work, combining them in the following theorem:

On the Hypothesis of the Right Angle and on the Hypothesis of the Obtuse Angle, a line perpendicular to a given straight line and a line cutting it at an acute angle intersect each other.

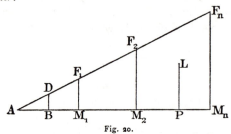

Fig. 20.

Let (Fig. 20) LP and AD be two straight lines of which the one is perpendicular to AP, and the other is inclined to AP at an acute angle DAP.

After cutting off in succession equal segments AD, DF_1, upon AD, draw the perpendiculars DB and F_1M_1 upon the line AP.

From Lemma III. above, we have

$$BM_1 \gtreqqless AB,$$

or $\qquad\qquad AM_1 \gtreqqless 2\ AB,$

on the two hypotheses.

Now cut off F_1F_2 equal to AF_1, from AF_1 produced, and let M_2 be the foot of the perpendicular from F_2 upon AP. Then we have

$$AM_2 \gtreqqless 2\ AM_1,$$

and thus

$$AM_2 \gtreqqless 2^2\ AB.$$

This process can be repeated as often as we please.

In this way we would obtain a point F_n upon the line AD such that its projection upon the line AP would determine a segment AM_n satisfying the relation

$$AM^n \gtreqless 2^n AB.$$

But if n is taken sufficiently great, [by the *Postulate of Archimedes* [1]] we would have

$$2^n AB > AP,$$

and therefore

$$AM_n > AP.$$

Therefore the point P lies upon the side AM_n of the right-angled triangle $AM_n F_n$. The perpendicular PL cannot intersect the other side of this triangle; therefore it cuts the hypotenuse.[2] *Q. E. D.*

It is now possible to prove the following theorem:

The Fifth Postulate is true on the Hypothesis of the Right Angle and on the Hypothesis of the Obtuse Angle [Prop. XIII.].

Let (Fig. 21) AB, CD be two straight lines cut by the line AC.

Let us suppose that

$$\sphericalangle BAC + \sphericalangle ACD < 2 \text{ right angles.}$$

Then one of the angles BAC, ACD, for example the first, will be acute.

From C draw the perpendicular CH upon AB. In the triangle ACH, from the hypotheses which have been made, we shall have

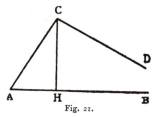

Fig. 21.

$$\sphericalangle A + \sphericalangle C + \sphericalangle H \gtreqless 2 \text{ right angles.}$$

---

[1] The *Postulate of Archimedes*, of which use is here made, includes implicitly the infinity of the straight line.

[2] The method followed by SACCHERI in proving this theorem is practically the same as that of NASÎR-EDDÎN. However Nasîr-Eddîn only deals with the *Hypothesis of the Right Angle*, as he had formerly shown that the sum of the angles of a triangle is equal to two right angles. It is right to remember that SACCHERI was familiar with and had criticised the work of the Arabian Geometer.

But we have assumed that

$$\measuredangle\, BAC + \measuredangle\, ACD < 2 \text{ right angles.}$$

These two results show that

$$\measuredangle\, AHC > \measuredangle\, HCD.$$

Thus the angle *HCD* must be acute, as *H* is a right angle. It follows from Prop. XI., XII. that the lines *AB* and *CD* intersect.[1]

This result allows SACCHERI to conclude that *the Hypothesis of the Obtuse Angle is false* [Prop. XIV.]. In fact, on this hypothesis EUCLID's Postulate holds [Prop. XIII.], and consequently, the usual theorems which are deduced from this postulate also hold. Thus the sum of the angles of the fundamental quadrilateral is equal to four right angles, so that the *Hypothesis of the Right Angle is true.*[2]

§ 16. But SACCHERI wishes to prove that the Fifth Postulate is true in every case. He thus sets himself to destroy *the Hypothesis of the Acute Angle.*

To begin with he shows that *on this hypothesis, a straight line being given, there can be drawn a perpendicular to it and a line cutting it at an acute angle, which do not intersect each other* [Prop. XVII.].

To construct these lines, let *ABC* (Fig. 22) be a triangle of which the angle *C* is a right angle. At *B* draw *BD* making the angle *ABD* equal to the angle *BAC*. Then, *on the*

---

[1] This proof is also found in the work of NASÎR-EDDÎN, which evidently inspired the investigations of SACCHERI.

[2] It should be noted that in this demonstration SACCHERI makes use of the special type of argument of which we spoke in § 11. In fact, from the assumption that the *Hypothesis of the Obtuse Angle* is true, we arrive at the conclusion that the *Hypothesis of the Right Angle* is true. This is a characteristic form taken in such cases by the ordinary reductio ad absurdum argument.

Hypothesis of the Acute Angle, the angle *CBD* is acute, and of the two lines *CA, BD*, which do not meet [Bk. I. 27], one makes a right angle with *BC*.

In what follows we consider only *the Hypothesis of the Acute Angle*.

Let (Fig. 23) *a, b* be two straight lines in the same plane which do not meet.

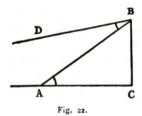

Fig. 22.

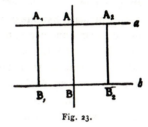

Fig. 23.

From the points A_1, A_2, on *a* draw perpendiculars A_1B_1, A_2B_2 to *b*.

The angles A_1, A_2 of the quadrilateral thus obtained can be

(i) one right, and one acute:

(ii) both acute:

(iii) one acute and one obtuse.

In the first case, there exists already a common perpendicular to the two lines *a, b*.

In the second case, we can prove the existence of such a common perpendicular by using the idea of continuity [SACCHERI, Prop. XXII.]. In fact, if the straight line A_1B_1 is moved continuously, while kept perpendicular to *b*, until it reaches the position A_2B_2, the angle $B_1A_1A_2$ starts as an acute angle and increases until it becomes an obtuse angle. There must be an intermediate position *AB* in which the angle BAA_2 is a right angle. Then *AB* is the common perpendicular to the two lines *a; b*.

In the third case, the lines *a, b* do not have a common

perpendicular, or, if such exists, it does not fall between B_1 and B_2.

Evidently there will be no such perpendicular if, for all the points A_r situated upon a, and on the same side of A_1, the quadrilateral $B_1A_1A_rB_r$ has always an obtuse angle at A_r.

With this *hypothesis* of the existence of two coplanar straight lines which do not intersect, and have no common perpendicular, SACCHERI proves that such lines always approach nearer and nearer to each other [Prop. XXIII.], and that their distance apart finally becomes smaller than any segment, taken as small as we please [Prop. XXV.]. In other words, if there are two coplanar straight lines, which do not cut each other, and have no common perpendicular, then these lines must be *asymptotic* to each other.[1]

To prove that such asymptotic lines effectively exist, SACCHERI proceeds as follows:—[2]

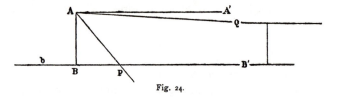

Fig. 24.

Among the lines of the pencil through A, coplanar with the line b, there exist lines which cut b, as, e. g., the line AB perpendicular to b; and lines which have a common

---

[1] With this result the question raised by the Greeks, as to the possibility of asymptotic lines in the same plane, is answered in the affirmative. Cf. p. 3.

[2] The statement of SACCHERI's argument upon the asymptotic lines differs in this edition from that given in the Italian and German editions. The changes introduced were suggested to me by some remarks of Professor CARSLAW.

perpendicular with b, as, e. g., the line AA' perpendicular to AB [cf. Fig. 24].

If AP cuts b, every other line of the pencil, which makes a smaller angle with AB than the acute angle BAP, also cuts b. On the other hand, if the line AQ, different from AA', has a common perpendicular with b, every other line, which makes with AB a larger acute angle than the angle BAQ, has a common perpendicular with b [cf. § 39, case (ii).]

Also it is clear that, if we take the lines of the pencil through A, from the ray AB towards the ray AA', we shall not find, among those which cut b, any line which is the last line of that set. In other words, the angles BAP, which the lines AP, cutting b, make with AB, have an *upper limit*, the angle BAX, such that the line AX does not cut b.

Then SACCHERI proves [Prop. XXX.] that, if we start with AA' and proceed in the pencil through A in the direction opposite to that just taken, we shall not find any last line in the set of lines which have a common perpendicular with b: that is to say, the angles BAQ, where AQ has a common perpendicular with b, have a *lower limit*, the angle BAY, such that the line AY does not cut b and has not a common perpendicular with b.

It follows that AY is a line asymptotic to b.

Further SACCHERI proves that the two lines AX and AY coincide [Prop. XXXII.]. His argument depends upon the consideration of points at infinity; and it is better to substitute for it another, founded on his Prop. XXI., viz., *On the Hypothesis of the Right Angle, and on that of the Acute Angle, the distance of a point on one of the lines containing an angle from the other bounding line increases indefinitely as this point moves further and further along the line.*

The suggested argument is as follows:

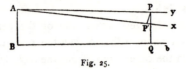

Fig. 25.

If AX [Fig. 25] does not coincide with AY, we can take a point P on AY, such that the perpendicular PP' from P to AX satisfies the inequality

(1) $PF' > AB$.   [Prop. XXI.]

On the other hand, if PQ is the perpendicular from P to b, the property of asymptotic lines [Prop. XXIII] shows that

$$AB > PQ.$$

But P is on the opposite side of AX from b.

Therefore $PQ > PP'$.

Combining this inequality with the preceding, we find that

$$AB > PP'.$$

which contradicts (1).

Hence AX coincides with AY.

We may sum up the preceding results in the following theorem:—

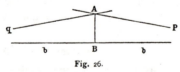

Fig. 26.

On the Hypothesis of the Acute Angle, there exist in the pencil of lines through A two lines p and q, asymptotic to b, one towards the right, and the other towards the left, which divide the pencil into two parts. The first of these consists of the lines which intersect b, and the second of those which have a common perpendicular with it.[1]

---

[1] In SACCHERI's work there will be found many other interesting theorems before he reaches this result. Of these the

§ 17. At this point SACCHERI attempts to come to a decision, trusting to intuition and to faith in the validity of the Fifth Postulate rather than to logic. To prove that *the Hypothesis of the Acute Angle is absolutely false, because it is repugnant to the nature of the straight line* [Prop. XXXIII.] he relies upon five Lemmas, spread over sixteen pages. In substance, however, his argument amounts to the *statement* that *if the Hypothesis of the Acute Angle were true, the lines p* (Fig. 26) *and b would have a common perpendicular at their common point at infinity, which is contrary to the nature of the straight line.* The so-called demonstration of SACCHERI is thus founded upon the extension to *infinity* of certain properties which are valid for figures at a finite distance.

However, SACCHERI is not satisfied with his reasoning and attempts to reach the wished-for proof by adopting anew the old idea of equidistance. It is not worth while to reproduce this second treatment as it does not contain anything of greater value than the discussions of his predecessors.

Still, though it failed in its aim, SACCHERI's work is of great importance. In it the most determined effort had been made on behalf of the Fifth Postulate; and the fact that he did not succeed in discovering any contradictions among the consequences of the *Hypothesis of the Acute Angle,* could not help suggesting the question, whether a consistent logical geometrical system could not be built upon this hypo-

---

following is noteworthy: *If two straight lines continually approach each other and their distance apart remains always greater than a given segment, then the Hypothesis of the Acute Angle is impossible.* Thus it follows that, if we postulate the absence of asymptotic straight lines, we must accept the truth of the Euclidean hypothesis.

thesis, and the Euclidean Postulate be impossible of demonstration.[1]

Johann Heinrich Lambert [1728—1777].

§ 18. It is difficult to say what influence SACCHERI's
work exercised upon the geometers of the 18th century.
However, it is probable that the Swiss mathematician
LAMBERT was familiar with it,[2] since in his *Theorie der Parallellinien* [1766] he quotes a dissertation by G. S. KLÜGEL
[1739—1812][3], where the work of the Italian geometer
is carefully analysed. LAMBERT's *Theorie der Parallellinien*
was published after the author's death, being edited by
J. BERNOULLI and C. F. HINDENBURG. It is divided into
three parts. The first part is of a critical and philosophical
nature. It deals with the two-fold question arising out of the
Fifth Postulate: whether it can be proved with the aid of
the preceding propositions only, or whether the help of some
other hypothesis is required. The second part is devoted to

---

[1] The publication of SACCHERI's work attracted considerable
attention. Mention is made of it in two Histories of Mathematics:
that of J. C. HEILBRONNER (Leipzig, 1742) and that of MONTUCLA
(Paris, 1758). Further it is carefully examined by G. S. KLÜGEL
in his dissertation noted below (Note (3)). Nevertheless it was
soon forgotten. Not till 1889 did E. BELTRAMI direct the attention
of geometers to it again in his Note: *Un precursore italiano
di Legendre e di Lobatschewsky*. Rend. Acc. Lincei (4), T. V. p. 441
—448. Thereafter SACCHERI's work was translated into English by
G. B. HALSTED (Amer. Math. Monthly, Vol. I. 1894 et seq.); into
German, by ENGEL and STÄCKEL (*Th. der P.* 1895); into Italian,
by G. Boccardini (Milan, Hoepli, 1904).

[2] Cf. SEGRE: *Congetture intorno alla influenza di Girolamo
Saccheri sulla formazione della geometria non euclidea*. Atti Acc.
Scienze di Torino, T. XXXVIII. (1903).

[3] *Conatuum praecipuorum theoriam parallelarum demonstrandi
recensio, quam publico examini submittent A. G. Kaestner et auctor
respondens G. S. Klügel*, (Göttingen, 1763).

the discussion of different attempts in which the Euclidean Postulate is reduced to very simple propositions, which however, in their turn, require to be proved. The third, and most important, part contains an investigation resembling that of SACCHERI, of which we now give a short summary.[1]

§ 19. LAMBERT's fundamental figure is a *quadrilateral with three right angles,* and three hypotheses are made as to the nature of the fourth angle. The first is the *Hypothesis of the Right Angle*; the second, the *Hypothesis of the Obtuse Angle*; and the third, the *Hypothesis of the Acute Angle.* Also in his treatment of these hypotheses the author does not depart far from SACCHERI's method.

The *first hypothesis* leads easily to the Euclidean system.

In rejecting the *second hypothesis*, LAMBERT relies upon a figure formed by two straight lines a, b, perpendicular to a third line AB (Fig. 27). From points $B, B_1, B_2, \ldots B_n$, taken in succession upon the line b, the perpendiculars, $BA, B_1A_1, B_2A_2, \ldots B_nA_n$ are drawn to the line a. He proves, in the first place, that these perpendiculars continually diminish, starting from the perpendicular BA. Next, that the difference between each and the one which succeeds it continually increases.

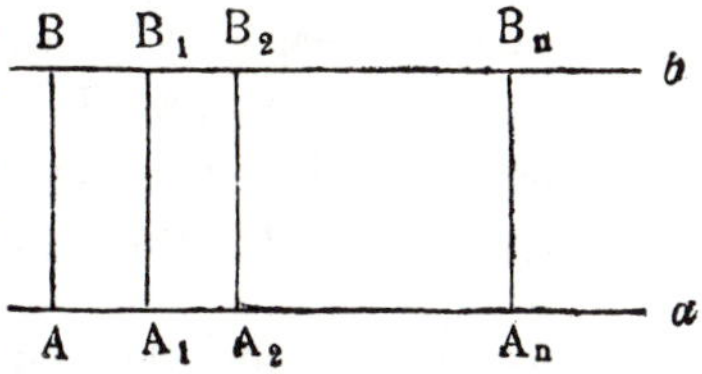

Fig. 27.

Therefore we have

$$BA - B_nA_n > n\,(BA - B_1A_1).$$

But, if n is taken sufficiently large, the second member

[1] Cf. Magazin für reine und angewandte Math., 2. Stück, p. 137—164. 3. Stück, p. 325—358, (1786). LAMBERT's work was again published by ENGEL and STÄCKEL (*Th. der P.*) p. 135—208, preceded by historical notes on the author.

of this inequality becomes as great as we please [*Postulate of Archimedes*] [1], whilst the first member is always less than *BA*. This contradiction allows LAMBERT to declare that the *second hypothesis* is false.

In examining the *third hypothesis,* LAMBERT again avails himself of the preceding figure. He proves that the perpendiculars $BA, B_1A_1, \ldots B_nA_n$ continually increase, and that at the same time the difference between each and the one which precedes it continually increases. As this result does not lead to contradictions, like SACCHERI he is compelled to carry his argument further. Then he finds, that, on the *third hypothesis* the sum of the angles of a triangle is less than two right angles; and going a step further than SACCHERI, he discovers that the *defect of a polygon,* that is, the difference between $2 \ (n-2)$ right angles and the sum of its angles, is *proportional to the area of the polygon.* This result can be obtained more easily by observing that both the area and the defect of a polygon, which is the sum of several others, are, respectively, the sum of the areas and of the defects of the polygons of which it is composed. [2]

§ 20. Another remarkable discovery made by LAMBERT has reference to the measurement of geometrical magnitudes. It consists precisely in this, that, whilst in the ordinary geometry only a *relative* meaning attaches to the choice of a

---

[1] The *Postulate of Archimedes* is again used here in a form which assumes the infinity of the straight line (cf. SACCHERI, Note p. 37).

[2] It is right to point out that in the *Hypothesis of the Acute Angle* SACCHERI had already met the *defect* here referred to, and also noted implicitly that a quadrilateral, made up of several others, has for its *defect* the sum of the *defects* of its parts (Prop. XXV). However he did not draw any conclusion from this as to the area being proportional to the *defect.*

particular unit in the measurement of lines, in the geometry founded upon the *third hypothesis*, we can attach to it an *absolute* meaning.

First of all we must explain the distinction, which is here introduced, between *absolute* and *relative*. In many questions it happens that the elements, supposed given, can be divided into two groups, so that those of the *first group* remain fixed, right through the argument, while those of the *second group* may vary in a number of possible cases. When this happens, the explicit reference to the data of the first group is often omitted. All that depends upon the varying data is considered *relative;* all that depends upon the fixed data is *absolute*.

For example, in the theory of the *Domain of Rationality*, the data of the *second group* [the variable data] are taken as certain simple irrationalities [constituting a *base*], and the *first group* consists simply of unity [1], which is often passed over in silence as it is common to all domains. In speaking of a number, we say that it is *rational relatively* to a given base, if it belongs to the domain of rationality defined by that base. We say that it is *rational absolutely*, if it is proved to be rational with respect to the base 1, which is common to all domains.

Passing to Geometry, we observe that in every actual problem, we generally take certain figures as given and therefore the magnitudes of their parts. In addition to these variable data [of the *second group*], which can be chosen in an arbitrary manner, there is always implicitly assumed the presence of the fundamental figures, straight lines, planes, pencils, etc. [fixed data or of the *first group*]. Thus, every construction, every measurement, every property of any figure ought to be held as *relative*, if it is essentially relative to the variable data. It ought, on the other hand, to be spoken of as *absolute*, if it is relative only to the fixed data.

[the fundamental figures], or, if, being enunciated in terms of the variable data, it only appears to depend upon them, so that it remains fixed when these vary.

In this sense it is clear that in ordinary geometry the measurement of lines has necessarily a relative meaning. Indeed the existence of similar figures does not allow us in any way to individualize the size of a line in terms of fundamental figures [straight line, pencil, etc.].

For an angle on the other hand, we can choose a method of measurement which expresses one of its absolute properties. It is sufficient to take its ratio to the angle of a complete revolution, that is, to the entire pencil, this being one of the fundamental figures.

We return now to LAMBERT and his geometry corresponding to the *third hypothesis*. He observed that with every segment we can associate a definite angle, which can easily be constructed. From this it follows that every segment is brought into correspondence with the fundamental figure [the pencil]. Therefore, in the new [hypothetical] geometry, we are entitled to ascribe an absolute meaning also to the measurement of segments.

To show in the simplest way how to every segment we can find a corresponding angle, and thus obtain an absolute numerical measurement of lines, let us imagine an equilateral triangle constructed upon every segment. We are able to associate with every segment the angle of the triangle corresponding to it and then the measure of this angle. Thus there exists a one-one correspondence between segments and the angles comprised between certain limits.

But the numerical representation of segments thus obtained does not enjoy the *distributive property* which belongs to *lengths*. On taking the sum of two segments, we do not obtain the sum of the corresponding angles. However, a function of the angle, possessing this property, can be obtained, and we can associate with the segment, not the said angle, but this function of the angle. For every value of the angle between certain limits, such a function gives an *absolute measure* of segments. The *absolute unit* of length is that segment for which this function takes the value 1.

Now if a certain function of the angle is distributive in the sense just indicated, the product of this function and an arbitrary constant also possesses that property. It is therefore clear that we can always choose this constant so that the absolute unit segment shall be that segment which corresponds to any assigned angle: e. g., 45°. The possibility of constructing the absolute unit segment, given the angle, depends upon the solution of the following problem:

To construct, on the Hypothesis of the Acute Angle, an equilateral triangle with a given defect.

So far as regards the absolute measure of the areas of polygons, we remark that it is given at once by the defect of the polygons. We can also assign an absolute measure for polyhedrons.

But with our intuition of space the absolute measure of all these geometrical magnitudes seems to us impossible. Hence *if we deny the existence of an absolute unit for segments, we can, with Lambert, reject the third hypothesis.*

§ 21. As LAMBERT realized the arbitrary nature of this statement, let it not be supposed that he believed that he had in this way proved the Fifth Postulate.

To obtain the wished-for proof, he proceeds with his investigation of the consequences of the *third hypothesis*, but he only succeeds in transforming his question into others equally difficult to answer.

Other very interesting points are contained in the *Theorie der Parallellinien*, for example, the close resemblance to spherical geometry[1] of the plane geometry which would hold, if the *second hypothesis* were valid, and the remark that spherical geometry is independent of the Parallel Postulate. Further, referring to the *third hypothesis*, he made the following acute and original observation: *From this I should almost conclude that the third hypothesis would occur in the case of an imaginary sphere.*

He was perhaps brought to this way of looking at the question by the formula $(A + B + C - \pi)\, r^2$, which expresses the area of a spherical triangle. If in this we write for the radius r, the imaginary radius $\sqrt{-1}\, r$ we obtain

$$r^2\,[\pi - A - B - C];$$

that is, the formula for the area of a plane triangle on LAMBERT's *third hypothesis*.[2]

§ 22. LAMBERT thus left the question in suspense. Indeed the fact that he did not publish his investigation allows us to conjecture that he may have discovered another way of regarding the subject.

Further, it should be remarked that, from the general want of success of these attempts, the conviction began to be formed in the second half of the 18th Century that it would be necessary to admit the Euclidean Postulate, or some other equivalent postulate, without proof.

In Germany, where the writings upon the question followed closely upon each other, this conviction had already assumed a fairly definite form. We recognize it in A. G. KÄSTNER,[3] a well-known student of the theory of parallels, and in his pupil, G. S. KLÜGEL, author of the

---

[1] In fact, in Spherical Geometry the sum of the angles of a quadrilateral is greater than four right angles, etc.

[2] Cf. ENGEL u. STÄCKEL; *Th. der P.* p. 146.

[3] For some information about KÄSTNER, cf. ENGEL u. STÄCKEL; *Th. der P.* p. 139—141.

valuable criticism of the most celebrated attempts to demonstrate the Fifth Postulate, referred to on p. 44 [note 3].
In this work KLÜGEL finds each of the proposed proofs
insufficient and suggests the possibility of non-intersecting
straight lines being divergent [*Möglich wäre es freilich, daß
Gerade, die sich nihct schneiden, voneinander abweichen*]. He
adds that the apparent contradiction which this presents is
not the result of a rigorous proof, nor a consequence of the
definitions of straight lines and curves, but rather something
derived from experience and the judgment of our senses.
[*Daß so etwas widersinnig ist, wissen wir nicht infolge strenger
Schlüsse oder vermöge deutlicher Begriffe von der geraden und
der krummen Linie, vielmehr durch die Erfahrung und durch
das Urteil unserer Augen*].

The investigations of SACCHERI and LAMBERT tend to
confirm KLÜGEL's opinion, but they cannot be held to be
a proof of the impossibility of demonstrating the Euclidean
hypothesis. Neither would a proof be reached if we proceeded along the way opened by these two geometers, and deduced any number of other propositions, not contradicting
the fundamendal theorems of geometry.

Nevertheless that one should go forward on this path,
without SACCHERI's presupposition that contradictions would
be found there, constitutes historically the decisive step in the
discovery that EUCLID's Postulate could not be proved, and
in the creation of the Non-Euclidean geometries.

But from the work of SACCHERI and LAMBERT to that of
LOBATSCHEWSKY and BOLYAI, which is based upon the above
idea, more than half a century had still to pass!

The French Geometers towards the End of the 18th Century.

§ 23. The critical study of the theory of parallels,
which had already led to results of great interest in Italy and

Germany, also made a remarkable advance in France towards the end of the 18th Century and the beginning of the 19th.

D'ALEMBERT [1717—1783], in one of his articles on geometry, states that 'La definition et les propriétés de la ligne droite, ainsi que des lignes parallèles sont l'écueil et pour ainsi dire le scandale des éléments de Géométrie.'[1] He holds that with a good definition of the straight line both difficulties ought to be avoided. He proposes to define a parallel to a given straight line as any other coplanar straight line, which joins two points which are on the same side of and equally distant from the given line. This definition allows parallel lines to be constructed immediately. However it would still be necessary to show that these parallels are equidistant. This theorem was offered, almost as a challenge, by D'ALEMBERT to his contemporaries.

§ 24. DE MORGAN, in his Budget of Paradoxes[2], relates that LAGRANGE [1736—1813], towards the end of his life, wrote a memoir on parallels. Having presented it to the French Academy, he broke off his reading of it with the exclamation: 'Il faut que j'y songe encore!' and he withdrew the MSS.

Further HOÜEL states that LAGRANGE, in conversation with BIOT, affirmed the independence of Spherical Trigonometry from EUCLID's Postulate.[3] In confirmation of this statement it should be added that LAGRANGE had made a special study of Spherical Trigonometry,[4] and that he inspired,

---

[1] Cf. D'ALEMBERT: *Mélanges de Litterature, d'Histoire, et de Philosophie*, T. V. § 11 (1759). Also: *Encyclopédie Méthodique Mathématique;* T. II. p. 519, Article: Parallèles (1785).

[2] A. DE MORGAN: *A Budget of Paradoxes*, p. 173. (London, 1872).

[3] Cf. J. HOÜEL: *Essai critique sur les principes fondamentaux de la géométrie élémentaire*, p. 84, Note (Paris, G. VILLARS, 1883).

[4] Cf. Miscellanea Taurinensia, T. II. p. 299—322 (1760—61).

if he did not write, a memoir '*Sur les principes fondamentaux de la Mécanique* [1760—1]'[1], in which FONCENEX discussed a question of independence, analogous to that above noted for Spherical Trigonometry. In fact, FONCENEX shows that the analytical law of the Composition of Forces acting at a point does not depend on the Fifth Postulate, nor upon any other which is equivalent to it.[2]

§ 25. The principle of similarity, as a fundamental notion, had been already employed by WALLIS in 1663 [cf. § 9]. It reappears at the beginning of the 19th Century, supported by the authority of two famous geometers: L. N. M. CARNOT [1753—1823] and LAPLACE [1749—1827].

In a Note [p. 481] to his *Géométrie de Position* [1803] CARNOT affirms that the theory of parallels is allied to the principle of similarity, the evidence for which is almost on the same plane as that for *equality*, and that, if this idea is once admitted, it is easy to establish the said theory rigorously.

LAPLACE [1824] observes that NEWTON's Law [the Law of Gravitation], by its simplicity, by its generality and by the confirmation which it finds in the phenomena of nature, must be regarded as rigorous. He then points out that one of its most remarkable properties is that, if the dimensions of all the bodies of the universe, their distances from each other, and their velocities, were to decrease proportionally, the heavenly bodies would describe curves exactly similar to those which they now describe, so that the universe, reduced step by step to the smallest imaginable space, would always present the same phenomena to its observers. These phenomena, he continues, are independent of the dimensions of the universe, so that the simplicity of the laws of nature only allows the observer to recognise their ratios. Referring again to this

---

[1] Cf. LAGRANGE: *Oeuvres*, T. VII. p. 331—363.
[2] Cf. Chapter VI.

astronomical conception of space, he adds in a Note: 'The attempts of geometers to prove EUCLID's Postulate on Parallels have been up till now futile. However no one can doubt this postulate and the theorems which EUCLID deduced from it. Thus the notion of space includes a special property, self-evident, without which the properties of parallels cannot be rigorously established. The idea of a bounded region, e. g., the circle, contains nothing which depends on its absolute magnitude. But if we imagine its radius to diminish, we are brought without fail to the diminution in the same ratio of its circumference and the sides of all the inscribed figures. This proportionality appears to me a more natural postulate than that of EUCLID, and it is worthy of note that it is discovered afresh in the results of the theory of universal gravitation.' [1]

§ 26. Along with the preceding geometers, it is right also to mention J. B. FOURIER [1768—1830], for a discussion on the straight line which he carried on with MONGE. [2] To bring this discussion into line with the investigations on parallels, we need only go back to D'ALEMBERT's idea that the demonstration of the postulate can be connected with the definition of the straight line [cf. § 23].

Fourier, who regarded the distance between two points as a *prime notion*, proposed to define first the sphere; then the plane, as the locus of points equidistant from two given points; [3] then the straight line, as the locus of the points equidistant from three given points. This method

---

[1] Cf. LAPLACE. *Oeuvres*, T. VI. Livre, V. Ch. V. p. 472.

[2] Cf. *Séances de l'École normale:* Débats, T. I. p. 28—33 (1795). This discussion was reprinted in Mathésis. T. IX. p. 139 —141 (1883).

[3] This definition of the plane was given by LEIBNITZ about a century before. Cf. *Opuscules et fragments inédits*, edited by L. COUTURAT, p. 554—5. (Paris, Alcan, 1903).

of presenting the problem of the foundations of geometry agrees with the opinions adopted at a later date by other geometers, who made a special study of the question of parallels [W. BOLYAI, N. LOBATSCHEWSKY, DE TILLY]. In this sense the discussion between FOURIER and MONGE finds a place among the earliest documents which refer to *Non-Euclidean geometry*.[1]

Adrien Marie Legendre [1752—1833].

§ 27. The preceding geometers confined themselves to pointing out difficulties and to stating their opinions upon the Postulate. Legendre, on the other hand, attempted to transform it into a theorem. His investigations, scattered among the different editions of his *Eléments de Géométrie* [1794—1823], are brought together in his *Refléxions sur différentes manières de démontrer la théorie des parallèles ou le théorème sur la somme des trois angles du triangle*. [Mém. Ac. Sc., Paris, T. XIII. 1833.]

In the most interesting of his attempts, LEGENDRE, like SACCHERI, approaches the question from the side of the sum of the angles of a triangle, which sum he wishes to prove equal to two right angles.

With this end in view, at the commencement of his work he succeeds in rejecting SACCHERI's *Hypothesis of the Obtuse Angle*, since he establishes *that the sum of the angles of any triangle is either less than* [*Hypothesis of the Acute Angle*] *or equal to* [*Hypothesis of the Right Angle*] *two right angles*.

We reproduce a neat and simple proof which he gives of this theorem:

Let n equal segments A_1A_2, A_2A_3, ... A_nA_{n+1} be taken

---

[1] To this we add that later memoirs and investigations showed that Fourier's definition also fails to build up the Euclidean theory of parallels, without the help of the Fifth Postulate, or some other equivalent to it.

one after the other on a straight line [Fig. 28]. On the same
side of the line let n equal triangles be constructed, having
for their third angular points $B_1 B_2 \ldots B_n$. The segments
$B_1 B_2$, $B_2 B_3$, $\ldots B_{n-1} B_n$, which join these vertices, are equal
and can be taken as the bases of n equal triangles, $B_1 A_2 B_2$,

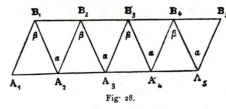

Fig· 28.

$B_2 A_3 B_3$, $\ldots B_{n-1}$
$A_n B_n$. The figure
is completed by
adding the triangle
$B_n \ A_{n+1} \ B_{n+1}$,
which is equal to
the others.

Let the angle B_1 of the triangle $A_1 B_1 A_2$ be denoted by
β, and the angle A_2 of the consecutive triangle by α.

Then $\beta \leqq \alpha$.

In fact, if $\beta > \alpha$, by comparing the two triangles $A_1 B_1 A_2$
and $B_1 A_2 B_2$, which have two equal sides, we would deduce

$$A_1 A_2 > B_1 B_2.$$

Further, since the broken line $A_1 B_1 B_2 \ldots B_{n+1} A_{n+1}$
is greater than the segment $A_1 A_{n+1}$,

$$A_1 B_1 + n. \ B_1 B_2 + A_{n+1} \ B_{n+1} > n. \ A_1 A_2,$$
$$\text{i. e., } 2 \ A_1 B_1 > n(A_1 A_2 - B_1 B_2).$$

But if n is taken sufficiently great, this inequality contradicts the *Postulate of Archimedes*.

Therefore $A_1 A_2$ is not greater than $B_1 B_2$,
and it follows that it is impossible that $\beta > \alpha$.

Thus we have $\beta \leqq \alpha$.

From this it readily follows that the sum of the angles of
the triangle $A_1 B_1 A_2$ less than or equal to two right angles.

This theorem is usually, but mistakenly, called *Legendre's
First Theorem*. We say mistakenly, because SACCHERI had
already established this theorem almost a century earlier [cf
p. 38] when he proved that the *Hypothesis of the Obtuse
Angle* was false.

The theorem usually called *Legendre's Second Theorem* was also given by SACCHERI, and in a more general form [cf. p. 29]. It is as follows:

If the sum of the angles of a triangle is less than or equal to two right angles in only one triangle, it is respectively less than or equal to two right angles in every other triangle.

We do not repeat the demonstration of this theorem, as it does not differ materially from that of SACCHERI.

We shall rather show how LEGENDRE proves *that the sum of the three angles of a triangle is equal to two right angles.*

Suppose that in the triangle ABC [cf. Fig. 29]

$$\angle A + \angle B + \angle C < 2 \text{ right angles.}$$

A point D being taken on AB, the transversal DE is drawn, making the angle ADE equal to the angle B. In the quadrilateral $DBCE$ the sum of the angles is less than 4 right angles.

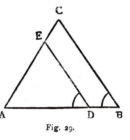

Fig. 29.

Therefore $\angle AED > \angle ACB$. The angle E of the triangle ADE is then a perfectly definite [decreasing] function of the side AD: or, what amounts to the same thing, the length of the side AD is fully determined when we know the size (in right angles) of the angle E, and of the two fixed angles A, B.

But this result Legendre holds to be absurd, since the length of a line has not a meaning, unless one knows the unit of length to which it is referred, and the nature of the question does not indicate this unit in any way.

In this way the hypothesis

$$\angle A + \angle B + \angle C < 2 \text{ right angles}$$

is rejected, and consequently we have

$$\angle A + \angle B + \angle C = 2 \text{ right angles.}$$

Also from this equality the proof of *Euclid's Postulate* follows easily.

LEGENDRE's method is thus based upon LAMBERT's postulate, which denies the existence of an *absolute unit* segment.

§ 28. In another demonstration LEGENDRE makes use of the hypothesis:

From any point whatever, taken within an angle, we can always draw a straight line which will cut the two arms of the angle.[1]

He proceeds as follows:

Let *ABC* be a triangle, in which, if possible, the sum of the angles is less than two right angles.

Let 2 right angles— $\sphericalangle A$ — $\sphericalangle B$ — $\sphericalangle C = \alpha$ [the *defect*].

Find the point *A'*, symmetrical to *A*, with respect to the side *BC*. [cf. Fig. 30.]

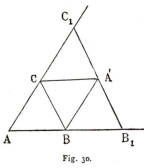

Fig. 30.

The defect of the new triangle *BCA'* is also α. In virtue of the hypothesis enunciated above, draw through *A'* a transversal meeting the arms of the angle *A* in B_1 and C_1. It can easily be shown that the defect of the triangle AB_1C_1 is the sum of the defects of the four triangles of which it is composed. [cf. also LAMBERT p. 46.]

Thus this defect is greater than 2 α.

Starting now with the triangle AB_1C_1 and repeating the same construction, we get a new triangle whose defect is greater than 4 α.

[1] J. F. LORENZ had already used this hypothesis for the same purpose. Cf. *Grundriß der reinen und angewandten Mathematik.* (Helmstedt, 1791).

After n operations of this kind a triangle will have been constructed whose defect is greater than $2^n \alpha$.

But for n sufficiently great, this defect, $2^n \alpha$, must be greater than 2 right angles [*Postulate of Archimedes*], which is absurd.

It follows that $\alpha = o$, and $\measuredangle A + \measuredangle B + \measuredangle C = 2$ right angles.

This demonstration is founded upon the *Postulate of Archimedes*. We shall now show how we could avoid using this postulate [cf. Fig. 31].

Let AB and HK be two straight lines, of which AB makes an acute angle, and HK a right angle, with AH.

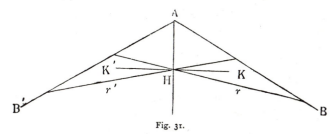

Fig. 31.

Draw the straight line AB' symmetrical to AB with regard to AH. Through the point H there passes, in virtue of LEGENDRE's hypothesis, a line r which cuts the two arms of the angle BAB'. If this line is different from HK, then also the line r', symmetrical to it with respect to AH, enjoys the same property of intersecting the arms of the angle. It follows that the line HK also meets them.

Thus the line perpendicular to AH and a line making an acute angle with AH always meet.

From this result the ordinary theory of parallels follows, and $\measuredangle A + \measuredangle B + \measuredangle C = 2$ right angles.

In other demonstrations LEGENDRE adopts the methods of analysis and also makes an erroneous use of infinity.

By these very varied investigations LEGENDRE believed that he had finally removed the serious difficulties surrounding the foundations of geometry. In substance, however, he added nothing new to the material and to the results obtained by his predecessors. His greatest merit lies in the elegant and simple form which he was able to give to all his writings. For this reason they gained a wide circle of readers and helped greatly to increase the number of disciples of the new ideas, which at that time were beginning to be formed.

Wolfgang Bolyai [1775—1856].

§ 29. In this article we come to the work of the Hungarian geometer W. BOLYAI. His interest in the theory of parallels dates back to the time when he was a student at Göttingen [1796—99], and is probably due to the advice of KÄSTNER and of his friend, the young Professor of Astronomy, K. F. SEYFFER [1762—1822].

In 1804 he sent GAUSS, formerly one of his student friends at Göttingen, a *Theoria Parallelarum,* which contained an attempt at a proof of the existence of equidistant straight lines.[1] GAUSS showed that this proof was fallacious. BOLYAI however, did not on this account give up his study of Axiom XI., though he only succeeded in substituting for it others, more or less evident. In this way he came to doubt the possibility of a demonstration and to conceive the impossibility of *doing away with the Euclidean hypothesis.* He asserted that the results derived from the denial of Axiom XI could not contradict the principles of geometry, since the law of the intersection of two straight lines, in its usual

---

[1] The *Theoria Parallelarum* was written in Latin. A German translation by ENGEL and STÄCKEL appears in Math. Ann. Bd. XLIX. p. 168—205 (1897).

form, represents a new datum, independent of those which precede it.[1]

WOLFGANG brought together his writings on the principles of mathematics in the work: *Tentamen juventutem studiosam in elementa Matheseos* [1832—33]; and in particular his investigations on *Axiom XI.*, while in each attempt he pointed out the new hypothesis necessary to render the demonstration rigorous.

A remarkable *postulate* to which WOLFGANG reduces EUCLID's is the following:

Four points, not on a plane, always lie upon a sphere; or, what amounts to the same thing: *A circle can always be drawn through three points not on a straight line.*[2]

The Euclidean Postulate can be deduced from this as follows [cf. Fig. 32]:

Let *AA'*, *BB'* be two straight lines, one of them being perpendicular to *AB*, and the other inclined to it at an acute angle.

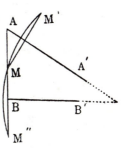

If we take a point *M* on the segment *AB* between *A* and *B*, and the points *M'M''* symmetrical to *M* with respect to the lines *BB'* and *AA'*, we obtain two points *M'*, *M''* not in the same straight line with *M*.

These three points *M*, *M'*, *M''* lie on the circumference of a circle. Also the lines *AA'*, *BB'* must intersect, since they both pass through the centre of this circle.

Fig. 32.

But from the fact that a line which is perpendicular to

[1] Cf. STÄCKEL: *Die Entdeckung der nichteuklidischen Geometrie durch J. Bolyai*, Math. u. Naturw. Ber. aus Ungarn, Bd. XVII. (1901).

[2] Cf. W. BOLYAI: *Kurzer Grundriss eines Versuchs etc.*, p. 46. (Maros Vásárhely, 1851).

another straight line and a line which cuts it at an acute angle intersect, it follows immediately that there can be only one parallel.

Friedrich Ludwig Wachter [1792—1817].

§ 30. When it had been seen that the Euclidean Postulate depends on the possibility of a circle being drawn through any three points not on a straight line, the idea at once suggested itself that the existence of such a circle should be established as a preliminary to any investigation of parallels.

An attempt in this direction was made by F. L. WACHTER.

WACHTER, a student under GAUSS in Göttingen [1809], and Professor of Mathematics in the Gymnasium of Dantzig, had made several attempts at the demonstration of the *Postulate*. He believed that he had been successful, first in a letter to GAUSS [Dec., 1816], and later, in a tract, printed at Dantzig in 1817.[1]

In this pamphlet he seeks to establish that given any four points in space, (not on a plane), a sphere will pass through them. He makes use of the following postulate:

Any four points of space fully determine a surface [the surface of four points], and two of these surfaces intersect in a single line, completely determined by three points.

There is no advantage in following the argument by means of which WACHTER seeks to prove that the *surface of four points* is a sphere, since he fails to give a precise definition of that surface in his tract. His deductions have thus only an intuitive character.

On the other hand a passage in his letter of 1816 deserves special notice. It was written after a conversation with GAUSS, when they had spoken of an *Anti-Euclidean Geometry*. In this letter he speaks of the surface to which a sphere tends

---

[1] *Demonstratio axiomatis geometrici in Euclideis undecimi.*

as its radius approaches infinity, a surface on the Euclidean hypothesis identical with a plane. He affirms that *even in the case of the Fifth Postulate being false, there would be a geometry on this surface identical with that of the ordinary plane.*

This statement is of the greatest importance as it contains one of the most remarkable results which hold in the system of geometry, corresponding to SACCHERI's Hypothesis of the Acute Angle [cf. LOBATSCHEWSKY, § 40].[1]

Bernhard Friedrich Thibaut [1775—1832].

§ 30 (bis). One other erroneous proof of the theorem that the sum of the angles of a triangle is equal to two right angles should be mentioned, since it has recently been revived in English textbooks, and to some extent received official sanction. It depends upon the idea of *direction*, and assumes that translation and rotation are independent operations. It is due to THIBAUT (*Grundriß der reinen Mathematik*, 2. Aufl., Göttingen, 1809). GAUSS refers to this "proof" in his correspondence with SCHUMACHER, and shows that it involves a proposition which not only needs proof, but is, in essence, the very proposition to be proved. THIBAUT argued as follows: [2]—

"Let *ABC* be any triangle whose sides are traversed in order from *A* along *AB*, *BC*, *CA*. While going from *A* to *B* we always gaze in the direction *ABb* (*AB* being produced to *b*), but do not turn round. On arriving at *B* we turn from the direction *Bb* by a rotation through the angle *bBC*, until we gaze in the direction *BCc*. Then we proceed in the direction *BCc* as far as *C*, where again we turn from *Cc* to *CAa* through the angle *cCA*; and at last arriving at *A*, we turn from the direction *Aa* to the first direction *AB* through the external angle *aAB*. This done, we have made a complete revolution,— just as if, standing at some point, we had turned completely round; and the measure of this rotation is 2 π. Hence the external angles of the triangle add up to 2 π, and the internal angles $A + B + C = \pi$. Q. E. D."

---

[1] With regard to WACHTER, cf. P. STÄCKEL: *Friedrich Ludwig Wachter, ein Beitrag zur Geschichte der nichteuklidischen Geometrie.* Math. Ann. Bd. LIV. p. 49—85. (1901). In this article are reprinted WACHTER's letters upon the subject and the tract of 1817 referred to above.

[2] [For further discussion of this "proof" see W. B. FRANKLAND's *Theories of Parallelism*, (Camb.Univ. Press, 1910), from which this version is taken, and HEATH's *Euclid*, Vol. I., p. 321.]

Chapter III.

The Founders of Non-Euclidean Geometry.

Carl Friederich Gauss [1777—1855].

§ 31. Twenty centuries of useless effort, and in particular the last unsuccessful investigations on the Fifth Postulate, convinced many of the geometers, who flourished about the beginning of last century, that the final settlement of the theory of parallels involved a problem whose solution was impossible. The Göttingen school had officially declared the necessity of admitting the Euclidean hypothesis. This view, expressed by KLÜGEL in his *Conatuum* [cf. p. 44] was accepted and supported by his teacher, A. G. KÄSTNER, then Professor in the University of Göttingen.[1]

Nevertheless keen interest was always taken in the subject; an interest which still continued to provide those who sought for a proof of the postulate with fruitless labour, and led finally to the discovery of new systems of geometry. These, founded like ordinary geometry on intuition, extend into a far wider field, freed from the principle embodied in the Euclidean Postulate.

How difficult was this advance towards the new order of ideas will be clear to any one who carries himself back to that period, and remembers the trend of the Kantian Philosophy, then predominant.

§ 32. GAUSS was the first to have a clear view of a geometry independent of the Fifth Postulate, but this re-

---

[1] Cf. ENGEL u. STÄCKEL: *Th. der P.* p. 139—142.

mained for quite fifty[1] years concealed in the mind of the great geometer, and was only revealed after the works of LOBATSCHEWSKY [1829—30] and J. BOLYAI [1832] appeared.

The documents which allow an approximate reconstruction of the lines of research followed by GAUSS in his work on parallels, are his correspondence with W. BOLYAI, OLBERS, SCHUMACHER, GERLING, TAURINUS and BESSEL [1799—1844]; two short articles in the *Gött. gelehrten Anzeigen* [1816, 1822]; and some notes found among his papers, [1831].[2]

Comparing the various passages in GAUSS's letters, we can fix the year 1792 as the date at which he began his '*Meditations*'.

The following portion of a letter to W. BOLYAI [Dec. 17, 1799] proves that GAUSS, like SACCHERI and LAMBERT before him, had attempted to prove the truth of Postulate V. by assuming it to be false.

'As for me, I have already made some progress in my work. However the path I have chosen does not lead at all to the goal which we seek, and which you assure me you have reached.[3] It seems rather to compel me to doubt the truth of geometry itself.

'It is true that I have come upon much which by most people would be held to constitute a proof: but in my eyes it proves as good as *nothing*. For example, if one could show that a rectilinear triangle is possible, whose area would be greater than any given area, then I would be ready to prove the whole of geometry absolutely rigorously.

'Most people would certainly let this stand as an Axiom; but I, no! It would, indeed, be possible that the area might

---

[1] [It would be more correct to say over thirty.]

[2] Cf. GAUSS, *Werke*, Bd. VIII. p. 157—268.

[3] It is to be remembered that W. BOLYAI was working at this subject in Göttingen and thought he had overcome his difficulties. Cf. § 29.

always remain below a certain limit, however far apart the three angular points of the triangle were taken.'

In 1804, replying to W. Bolyai on his *Theoria parallelarum*, he expresses the hope that the obstacles by which their investigations had been brought to a standstill would finally leave a way of advance open.[1]

From all this, Stäckel and Engel, who collected and verified Gauss's correspondence on this subject, come to the conclusion that the great geometer did not recognize the existence of a logically sound Non-Euclidean geometry by intuition or by a flash of genius: that, on the contrary, he had spent upon this subject many laborious hours before he had overcome the inherited prejudice against it.

Did Gauss, when he began his investigations, know the writings of Saccheri and Lambert? What influence did they exert upon his work? Segre, in his *Congetture*, already referred to [p. 44 note 2], remarks that both Gauss and W. Bolyai, while students at Göttingen, the former from 1795 —98, the later from 1796—99, were interested in the theory of parallels. It is therefore possible that, through Kästner and Seyffer, who were both deeply versed in this subject they had obtained knowledge both of the *Euclides ab omni naevo vindicatus* and of the *Theorie der Parallellinien*. But the dates of which we are certain, although they do not contradict this view, fail to confirm it absolutely.

§ 33. To this first period of Gauss's work, after 1813 there follows a second. Of it we obtain some knowledge chiefly from a few letters, one written by Wachter to Gauss [1816]; others sent by Gauss to Gerling [1819], Taurinus [1824] and Schumacher [1831]; and also from some notes found among Gauss's papers.

---

[1] [It should be noticed that these efforts were still directed towards proving the truth of Euclid's postulate.]

These documents show us that GAUSS, in this second period, had overcome his doubts, and proceeded with his development of the fundamental theorems of a new geometry, which he first calls *Anti-Euclidean* [cf.WACHTER's letter quoted on p. 62]; then *Astral Geometry* [following SCHWEIKART, cf. p. 76]; and finally, *Non-Euclidean* [cf. letter to SCHUMACHER]. Thus he became convinced that the Non-Euclidean Geometry did not in itself involve any contradiction, though at first sight some of its results had the appearance of paradoxes [letter to SCHUMACHER, July 12, 1831].

However GAUSS did not let any rumour of his opinions get abroad, being certain that he would be misunderstood. [He was afraid of the *clamour of the Boeotians*; letter to BESSEL, Jan. 27, 1829]. Only to a few trusted friends did he reveal something of his work. When circumstances compel him to write to TAURINUS [1824] on the subject, he begs him to keep silence as to the information which he imparted to him.

The notes found among GAUSS's papers contain two brief synopses of the new theory of parallels, and probably belong to the projected exposition of the *Non-Euclidean Geometry*, with regard to which he wrote to SCHUMACHER [on May 17, 1831]: 'In the last few weeks I have begun to put down a few of my own *Meditations*, which are already to some extent nearly 40 years old. These I had never put in writing, so that I have been compelled three or four times to go over the whole matter afresh in my head. Also I wished that it should not perish with me.'

§ **34.** GAUSS defines parallels as follows:[1]

If the coplanar straight lines AM, BN, do not intersect each other, while, on the other hand, every straight line through

---

[1] [In this section upon GAUSS's work on Parallels fuller use has been made of the material in his Collected Works (GAUSS, *Werke*, Bd. VIII, p. 202—9)].

A between AM and AB cuts BN, then AM is said to be parallel to BN [fig. 33].

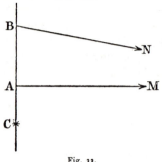

Fig. 33.

He supposes a straight line passing through A, to start from the position AB, and then to rotate continuously on the side towards which BN is drawn, till it reaches the position AC, in BA produced. This line begins by cutting BN and in the end it does not cut it. Thus there can be one and only one position, separating the lines which intersect BN from those which do not intersect it. This must be the *first* of the lines, which do not cut BN: and thus from our definition it is the parallel AM; since there can obviously be no *last* line of the set of lines which intersect BN.

It will be seen in what way this definition differs from EUCLID's. If EUCLID's Postulate is rejected, there could be different lines through A, on the side towards which BN is drawn, which would not cut BN. These lines would all be parallels to BN according to EUCLID's Definition. In GAUSS's definition only the first of these is said to be parallel to BN.

Proceeding with his argument GAUSS now points out that in his definition the starting points of the lines AM and BN are assumed, though the lines are supposed to be produced indefinitely in the directions of AM and BN.

I. He proceeds to show that *the parallelism of the line AM to the line BN is independent of the points A and B, provided the sense in which the lines are to be produced indefinitely remain the same.*

It is obvious that we would obtain the same parallel AM

if we kept A fixed and took instead of B another point B' on the line BN, or on that line produced backwards.

It remains to prove that if AM is parallel to BN for the point A, it is also the parallel to BN for any point upon AM, or upon AM produced backwards.

Instead of A [Fig. 34] take another starting point A' upon AM. Through A', between $A'B$ and $A'M$, draw the line $A'P$ in any direction.

Through Q, any point on $A'P$, between A' and P, draw the line AQ.

Then, from the definition, AQ must cut BN, so that it is clear QP must also cut BN.

Thus $AA'M$ is the first of the lines which do not cut BN, and $A'M$ is parallel to BN.

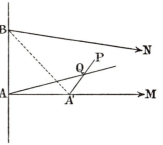

Fig. 34.

Again take the point A' upon AM produced backwards [Fig. 35].

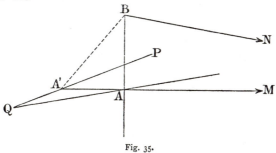

Fig. 35.

Draw through A', between $A'B$ and $A'M$, the line $A'P$ in any direction.

Produce $A'P$ backwards and upon it take any point Q. Then, by the definition, QA must cut BN, for example, in R. Therefore $A'P$ lies within the closed figure $A'ARB$, and must cut one of the four sides $A'A$, AR, RB, and BA'.

Obviously this must be the third side RB, and therefore $A'M$ is parallel to BN.

II. The Reciprocity of the Parallelism can also be established.

In other words, *if AM is parallel to BN, then BN is also parallel to AM.*

GAUSS proves this result as follows:

From any point B upon BN draw BA perpendicular to AM. Through B draw any line BN' between BA and BN.

At B, on the same side of AB as BN, make

$$\measuredangle ABC = {}^1/_2 \measuredangle N'BN.$$

There are two possible cases:

Case (i), when BC cuts AM [cf. Fig. 36].

Case (ii), when BC does not cut AM [cf. Fig. 37].

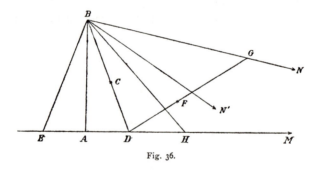

Fig. 36.

Case (i). Let BC cut AM in D. Take $AE = AD$, and join BE. Make $\measuredangle BDF = \measuredangle BED$.

Since AM is parallel to BN, DF must cut BM, for example, in G.

From EM cut off EH equal to DG.

Then, in the triangles BEH and BDG, it follows that

$$\angle EBH = \angle DBG.$$

Therefore $\angle EBD = \angle HBG.$

But $\angle EBD = \angle N'BN.$

Therefore BN' and BH coincide, and BN' must cut AM.

But BN' is *any* line through B, between BA and BN. Therefore BN is parallel to AM.

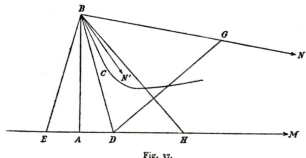

Fig. 37.

Case (ii). In this case let D be any arbitrary point upon AM. Then with the same argument as above,

$$\angle EBD = \angle GBH.$$

But $\angle ABD < \angle ABC.$

Therefore $\angle EBD < \angle N'BN.$

Therefore $\angle GBH < \angle N'BN.$

Therefore BN' must cut AM.

But BN' is any line through B, between BA and BN. Therefore BN is parallel to AM.

Thus in both cases we have proved that if AM is parallel to BN, then BN is parallel to AM.[1]

The next theorem proved by GAUSS in this synopsis is as follows:

[1 GAUSS's second proof of this theorem is given in the German translation. However it will be found that in it he assumes that BC cuts AM, and to prove this the argument used above is necessary.]

III. *If the line* (1) *is parallel to the line* (2) *and to the line* (3), *then* (2) *and* (3) *are parallel to each other.*

Case (i). Let the line (1) lie between (2) and (3) [cf. Fig. 38].

Let *A* and *B* be two points on (2) and (3), and let *AB* cut (1) in *C*.

Through *A* let an arbitrary line *AD* be drawn between

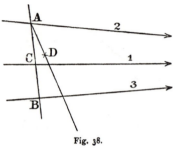

Fig. 38.

AB and (2). Then it must cut (1), and on being produced must also cut (3).

Since this holds for every line such as *AD*, (2) is parallel to (3).

Case (ii). Let the line (1) be outside both (2) and (3), and let (2) lie between (1) and (3) [cf. Fig. 39].

If (2) is not parallel to (3), through any point chosen at random upon (3), a line different from (3) can be drawn which is parallel to (2).

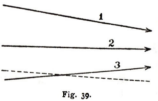

Fig. 39.

This, by Case (i), is also parallel to (1), which is absurd.

This short Note on Parallels closes with the theorem that *if two lines AM and BN are parallel, these lines produced backwards cannot meet.*

From all this it is evident that the parallelism of GAUSS means *parallelism in a given sense.* Indeed his definition of parallels deals with a line drawn from *A* on a definite side of the transversal *AB*: e. g., the *ray* drawn to the right, so that we might speak of *AM* as the parallel to *BN* towards the right. The parallel from *A* to *BN* towards the left is not necessarily *AM.* If it were, we would obtain the Euclidean hypothesis.

The two lines, in the third theorem, which are each parallel to a third line, are thus both parallels in the same sense (both left-hand, or both right-hand parallels).

In a second memorandum on parallels, GAUSS goes over the same ground, but adds the idea of Corresponding Points on two parallels *AA′, BB′*. *Two points A, B are said to correspond, when AB makes equal internal angles with the parallels on the same side* [cf. Fig. 40].

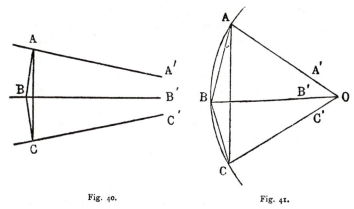

Fig. 40.   Fig. 41.

With regard to these Corresponding Points he states the following theorems:

(i) *If A, B are two corresponding points upon two parallels, and M is the middle point of AB, the line MN, perpendicular to AB, is parallel to the two given lines, and every point on the same side of MN as A is nearer A than B.*

(ii) *If A, B are two corresponding points upon the parallels* (1) *and* (2), *and A′, B′ two other corresponding points on the same lines, then AA′ = BB′, and conversely.*

(iii) *If A, B, C are three points on the parallels* (1), (2) *and* (3), *such that A and B, B and C, correspond, then A and C also correspond.*

The idea of Corresponding Points, when taken in connection with three lines of a pencil (that is, three concurrent lines [cf. Fig. 41] allows us to define the circle as *the locus of the points on the lines of a pencil which correspond to a given point.* But this locus can also be constructed when the lines of the pencil are parallel. In the Euclidean case the locus is a straight line: but putting aside the Euclidean hypothesis, the locus in question is a line, having many properties in common with the circle, but yet not itself a circle. Indeed *if any three points are taken upon it, a circle cannot be drawn through them.* This line can be regarded as the limiting case of a circle, when its radius becomes infinite. In the Non-Euclidean geometry of LOBATSCHEWSKY and BOLYAI, this locus plays a most important part, and we shall meet it there under the name of the Horocycle.[1]

This work GAUSS did not need to complete, for in 1832 he received from WOLFGANG BOLYAI a copy of the work of his son JOHANN on Absolute Geometry.

From letters before and after the date at which he interrupted his work, we know that GAUSS had discovered in his geometry an Absolute Unit of Length [cf. LAMBERT and LEGENDRE], and that a constant k appeared in his formulae, by means of which all the problems of the Non-Euclidean Geometry could be solved [letter to Taurinus, Nov. 8, 1824].

Speaking more fully of these matters in 1831 [letter to

---

1 [LOBATSCHEWSKY: *Grenzkreis, Courbe-limite* or *Horicycle.* BOLYAI; *Parazykl, L-linie.*

It is interesting to notice that GAUSS, even at this date, seems to have anticipated the importance of the Horocycle. The definition of Corresponding Points and the statement of their properties is evidently meant to form an introduction to the discussion of the properties of this curve, to which he seems to have given the name *Trope.*]

SCHUMACHER], he gave the length of the circumference of a circle of radius r in the form

$$\pi k \left(e^{\frac{r}{k}} - e^{-\frac{r}{k}} \right).$$

With regard to k, he says that, if we wish to make the new geometry agree with the facts of experience, we must suppose k infinitely great in comparison with all known measurements.

For $k = \infty$, GAUSS's expression takes the usual form for the perimeter of a circle.[1] The same remark holds for the whole of GAUSS's system of geometry. It contains EUCLID's system, as the limiting case, when $k = \infty$.[2]

Ferdinand Karl Schweikart [1780—1859].

§ 35. The investigations of the Professor of Jurisprudence, F. K. SCHWEIKART,[3] date from the same period as those of GAUSS, but are independent of them. In 1807 he published *Die Theorie der Parallellinien nebst dem Vorschlage ihrer Verbannung aus der Geometrie.* Contrary to what one might expect from its title, this work does not contain a treatment of parallels independent of the *Fifth Postulate*, but one based on the idea of the parallelogram.

But· at a later date, SCHWEIKART, having discovered a new order of ideas, developed a geometry independent of *Euclid's hypothesis*. When in Marburg in December, 1818, he handed the following memorandum to his colleague GERLING, asking him to communicate it to GAUSS and obtain his opinion upon it:

---

[1] To show this we need only use the exponential series.

[2] For other investigations by GAUSS, cf. Note on p. 90.

[3] He studied law at Marburg and from 1796—98 attended the lectures on Mathematics given in that University by Professor J. K. F. HAUFF, the author of various memoirs on parallels, cf. *Th. der P.* p. 243.

MEMORANDUM.

'There are two kinds of geometry—a geometry in the strict sense—the Euclidean; and an astral geometry [astra-lische Größenlehre].

'Triangles in the latter have the property that the sum of their three angles is not equal to two right angles.'

'This being assumed, we can prove rigorously:

a) That the sum of the three angles of a triangle is less than two right angles;

b) that the sum becomes ever less, the greater the area of the triangle;

c) that the altitude of an isosceles right-angled triangle continually grows, as the sides increase, but it can never become greater than a certain length, which I call the *Constant*.

Squares have, therefore, the following form [Fig. 42].

'If this Constant were *for us* the Radius of the Earth,

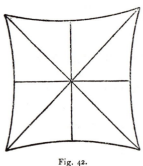

Fig. 42.

(so that every line drawn in the universe from one fixed star to another, distant 90° from the first, would be a tangent to the surface of the earth), it would be infinitely great in comparison with the spaces which occur in daily life.

'The Euclidean geometry holds only on the assumption that the Constant is infinite. Only in this case is it true that the three angles of every triangle are equal to two right angles: and this can easily be proved, as soon as we admit that the Constant is infinite.'[1]

SCHWEIKART's Astral Geometry and GAUSS's Non-Euclid-

ean Geometry exactly correspond to the systems of SAC-
CHERI and LAMBERT for the *Hypothesis of the Acute Angle.*
Indeed the contents of the above memorandum can be obtained directly from the theorems of SACCHERI, stated in
KLÜGEL's *Conatuum*, and from LAMBERT's Theorem on the
area of a triangle. Also since SCHWEIKART in his *Theorie* of
1807 mentions the works of the two latter authors, the direct
influence of LAMBERT, and, at least, the indirect influence of
SACCHERI upon his investigations are established.[2]

In March, 1819 GAUSS replied to GERLING with regard
to the Astral Geometry. He compliments SCHWEIKART, and
declares his agreement with all that the sheet of paper sent
to him contained. He adds that he had extended the Astral
Geometry so far that he could completely solve all its problems, if only SCHWEIKART's Constant were given. In conclusion, he gives the upper limit for the area of a triangle
in the form[3]

$$\frac{\pi\, CC}{[\log \text{hyp}\,(1 + \sqrt{2})]^2}\,.$$

SCHWEIKART did not publish his investigations.

Franz Adolf Taurinus [1794—1874].

§ 36. In addition to carrying on his own investigations
on parallels, SCHWEIKART had persuaded [1820] his nephew
TAURINUS to devote himself to the subject, calling his atten-

---

[1] Cf. GAUSS, *Werke*, Bd. VIII, p. 180—181.

[2] Cf. SEGRE's *Congetture*, cited above on p. 44.

[3] The constant which appears in this formula is SCHWEIKART's
Constant *C*, not GAUSS's constant *k*, in terms of which he expressed
the length of the circumference of a circle. (cf. p. 75). The two
constants are connected by the following equation:

$$k = \frac{C}{\log\,(1 + \sqrt{2})}.$$

tion to the *Astral Geometry*, and to GAUSS's favourable verdict upon it.

TAURINUS appears to have taken up the subject seriously for the first time in 1824, but with views very different from his uncle's. He was then convinced of the absolute truth of the Fifth Postulate, and always remained so, and he cherished the hope of being able to prove it. Failing in his first attempts, under the influence of GAUSS and SCHWEIKART, he again began the study of the question. In 1825 he published a *Theorie der Parallellinien*, containing a treatment of the subject on Non-Euclidean lines, the rejection of *the Hypothesis of the Obtuse Angle*, and some investigations resembling those of SACCHERI and LAMBERT on the *Hypothesis of the Acute Angle*. He found in this way SCHWEIKART's Constant, which he called a Parameter. He thought an absolute unit of length impossible, and concluded that all the systems, corresponding to the infinite number of values of the parameter, ought to hold simultaneously. But this, in its turn, led to considerations incompatible with his conception of space, and thus TAURINUS was led to reject the *Hypothesis of the Acute Angle* while recognising the *logical compatibility* of the propositions which followed from it.

In the next year TAURINUS published his *Geometriae Prima Elementa* [Cologne, 1826], in which he gave an improved version of his researches of 1825. This work concludes with a most important appendix, in which the author shows how a system of analytical geometry could be actually constructed on the *Hypothesis of the Acute Angle.*[1]

With this aim TAURINUS starts from the fundamental formula of Spherical Trigonometry—

---

[1] For the final influence of SACCHERI and LAMBERT upon TAURINUS, cf. SEGRE's *Congetture*, quoted above on p. 44.

$$\cos \frac{a}{k} = \cos \frac{b}{k} \ \cos \frac{c}{k} + \sin \frac{b}{k} \sin \frac{c}{k} \ \cos A,$$

In it he transforms the real radius k into the imaginary radius ik. Using the notation of the hyperbolic functions, we thus have

(1) $\qquad \cosh \dfrac{a}{k} = \cosh \dfrac{b}{k} \cosh \dfrac{c}{k} - \sinh \dfrac{b}{k} \sinh \dfrac{c}{k} \cos A.$

This is the fundamental formula of the *Logarithmic-Spherical Geometry* [*logarithmisch-sphärischen Geometrie*] of TAURINUS.

It is easy to show that in this geometry the sum of the angles of a triangle is less than 180°. For simplicity we take the case of an equilateral triangle, putting $a=b=c$ in (1).

Solving, for cos A, we obtain

(1*) $\qquad\qquad \cos A = \dfrac{\cosh \dfrac{a}{k}}{\cosh \dfrac{a}{k} + 1}.$

But sech $\dfrac{a}{k} < 1$.

Therefore cos $A > {}^1/_2$.

Thus A is less than 60°, and the sum of the angles of the triangle is less than 180°.

It is instructive to note, that, from (1*).

$$\underset{a \,=\, 0}{\text{Lt. }} (\cos A) = {}^1/_2.$$

So that in the limit when a becomes zero, A is equal to 60°. Therefore, in the *log.-spherical geometry*, the sum of the angles of a triangle tends to 180° when the sides tend to zero.

We may also note that from (1*)

$$\underset{k \,=\, \infty}{\text{Lt. }} (\cos A) = {}^1/_2;$$

so that in the limit when k is infinite, A is equal to 60°. Therefore, when the constant k tends to infinity, the angles of the equilateral triangle are each equal to 60°, as in the ordinary geometry.

More generally, using the exponential forms for the hyperbolic functions, it will be seen that in the limit when k is infinite (1) becomes

$$a^2 = b^2 + c^2 - 2bc \cos A,$$

the fundamental formula of Euclidean Plane Trigonometry.

§ 37. The second fundamental formula of Spherical Trigonometry,

$$\cos A = - \cos B \cos C + \sin B \sin C \cos \frac{a}{k},$$

by simply interchanging the cosine with the hyperbolic cosine, gives rise to the second fundamental formula of the *log.-spherical* geometry:

(2) $\cos A = - \cos B \cos C + \sin B \sin C \cosh \frac{a}{k}.$

For $A = 0$ and $C = 90°$, we have

(3) $$\cosh \frac{a}{k} = \frac{1}{\sin B}.$$

The triangle corresponding to this formula has one angle zero and the two sides containing it are infinite and parallel [asymptotic]. [Fig. 43.] The angle B, between the side which

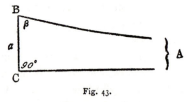

Fig. 43.

is parallel and the side which is perpendicular to CA, is seen from (3) to be a function of a. From this onward we can call it the *Angle of Parallelism* for the distance a [cf. LOBATSCHEWSKY, p. 87].

For $B = 45°$, the segment BC, which is given by (3), is SCHWEIKART's Constant [cf. p. 76]. Thus, denoting it by P,

$$\cosh \frac{P}{k} = \sqrt{2},$$

from which, solving for k, we have

$$k = \frac{P}{\log (1 + \sqrt{2})}.$$

This relation connecting the two constants P and k was given by TAURINUS. The constant k is the same as that employed by GAUSS [cf. p. 75] in finding the length of the circumference of a circle.

§ 38. TAURINUS deduced other important theorems in the *log.-spherical geometry* by further transformations of the formulæ of Spherical Trigonometry, replacing the real radius by an imaginary one.

For example, that the area of a triangle is proportional to its defect [LAMBERT, p. 46]:

that the superior limit of that area is

$$\frac{\pi P^2}{[\log (1+\sqrt{2})]^2} \text{ [GAUSS, p. 77]};$$

that the length of the circumference of a circle of radius r is

$$2 \pi k \sinh \frac{r}{k} \text{ [GAUSS, p. 75]};$$

that the area of a circle of radius r is

$$2 \pi k^2 (\cosh \frac{r}{k} - 1);$$

that the area of the surface of a sphere and its volume, are respectively

$$4 \pi k^2 \sinh^2 \frac{r}{k},$$

and $$2 \pi k^3 (\sinh \frac{r}{k} \cosh \frac{r}{k} - \frac{r}{k}).$$

We shall not devote more space to the different analyt-

ical developments, since a fuller discussion would cast no
fresh light upon the method. However we note that the
results of TAURINUS confirm the prophecy of LAMBERT on
the *Third Hypothesis* [cf. p. 50], since the formulæ of the
log.-spherical geometry, interpreted analytically, give the fundamental relations between the elements of a triangle traced
upon a sphere of imaginary radius.[1]

To this we add that TAURINUS in common with LAMBERT
recognized that Spherical Geometry corresponds exactly to
the system valid in the case of the *Hypothesis of the Obtuse
Angle:* further that the ordinary geometry forms a link between spherical geometry and the *log.-spherical geometry.*

Indeed, if the radius *k* passes continuously from the real
domain to the purely imaginary one, through infinity, we proceed from the spherical system to the *log.-spherical* system,
through the Euclidean.

Although TAURINUS, as we have already remarked, excluded the possibility that a *log.-spherical geometry* could be
valid on the plane, the theoretical interest, which it offers,
did not escape his notice. Calling the attention of geometers to his formulæ, he seemed to prophecy the existence

---

[1] At this stage it should be remarked that LAMBERT, simultaneously with his researches on parallels, was working at the trigonometrical functions with an imaginary argument, whose connection
with Non-Euclidean Geometry was brought to light by TAURINUS.
Perhaps LAMBERT recognised that the formulae of Spherical Trigonometry were still real, even when the real radius was changed
in a purely imaginary one. In this case his prophecy with regard
to the *Hypothesis of the Acute Angle* (cf. p. 50) would have a firm
foundation in his own work. However we have no authority for
the view that he had ever actually compared his investigations on
the trigonometrical functions with those on the theory of parallels.
Cf. P. STÄCKEL: *Bemerkungen zu Lamberts Theorie der Parallellinien.*
Biblioteca Math. p. 107—110. (1899).

of some concrete case in which they would find an interpretation.[1]

---

[1] The important service rendered by SCHWEIKART and TAURINUS towards the discovery of the Non-Euclidean Geometry was recognised and made known by ENGEL and STÄCKEL. In their *Th. der P.*, they devote a whole chapter to those authors, and quote the most important passages in TAURINUS' writings, besides some letters which passed between him, GAUSS and SCHWEIKART. Cf. STÄCKEL: *Franz Adolf Taurinus*, Abhandl. zur Geschichte der Math., IX, p. 397—427 (1899).

Chapter IV.

The Founders of Non-Euclidean Geometry (Contd.).

Nicolai Ivanovitsch Lobatschewsky [1793—1856].[1]

§ 39. LOBATSCHEWSKY studied mathematics at the University of Kasan under a German J. M. C. BARTELS [1769—1836], who was a friend and fellow countryman of GAUSS. He took his degree in 1813 and remained in the University, first as Assistant, and then as Professor. In the latter position he lectured upon mathematics in all its branches and also upon physics and astronomy.

As early as 1815 LOBATSCHEWSKY was working at parallels, and in a copy of his notes for his lectures [1815—17] several attempts at the proof of the Fifth Postulate, and some investigations resembling those of LEGENDRE have been found.

However it was only after 1823 that he had thought of the Imaginary Geometry. This may be inferred from the manuscript for his book on Elementary Geometry, where he says that we do not possess any proof of the Fifth Postulate, but that such a proof may be possible.[2]

---

[1] For historical and critical notes upon LOBATSCHEWSKY we refer once and for all to F. ENGEL's book: N. I. LOBATSCHEFSKIJ: *Zwei geometrische Abhandlungen aus dem Russischen übersetzt mit Anmerkungen und mit einer Biographie des Verfassers.* (Leipzig, Teubner, 1899).

[2] [This manuscript had been sent to St. Petersburg in 1823 to be published. However it was not printed, and it was dis-

Between 1823 and 1825 LOBATSCHEWSKY had turned his attention to a geometry independent of Euclid's hypothesis. The first fruit of his new studies is the *Exposition succincte des principes de la géométrie avec une démonstration rigoureuse du théorème des parallèles*, read on 12 [24] Feb., 1826, to the Physical Mathematical Section of the University of Kasan. In this "Lecture", the manuscript of which has not been discovered, LOBATSCHEWSKY explains the principles of a geometry, more general than the ordinary geometry, where two parallels to a given line can be drawn through a point, and where the sum of the angles of a triangle is less than two right angles [*The Hypothesis of the Acute Angle* of SACCHERI and LAMBERT].

Later, in 1829—30, he published a memoir *On the Principles of Geometry*,[1] containing the essential parts of the preceding "Lecture", and further applications of the new theory in analysis. In succession appeared the *Imaginary Geometry* [1835],[2] *New Principles of Geometry, with a Com-*

---

covered in the archives of the University of Kasan in 1898. It is clear from some other remarks in this work that he had made further advance in the subject since 1815—17. He was now convinced that all the first attempts at a proof of the Parallel Postulate were unsuccessful, and that the assumption that the angles of a triangle could depend only on the ratio of the sides and not upon their absolute lengths was unjustifiable (cf. ENGEL, loc. cit. p. 369—70).]

[1] Kasan Bulletin, (1829—1830). *Geometrical Works of Lobatschewsky* (Kasan 1883—1886), Vol. I p. 1—67. German translation by F. ENGEL p. 1—66 of the work referred to on the previous page.

Where the titles are given in English we refer to works published in Russian. The *Geometrical Works of Lobatschewsky* contain two parts; the first, the memoirs originally published in Russian; the second, those published in French or German. It will be seen below that of the works in Vol. I. several translations are now to be had.

[2] The Scientific Publications of the University of Kasan (1835). *Geometrical Works*, Vol. I, p. 71—120. German translation by

plete Theory of Parallels [1835—38][1], *the Applications of the Imaginary Geometry to Some Integrals* [1836][2], then the *Géométrie Imaginaire* [1837][3], and in 1840, a small book containing a summary of his work, *Geometrische Untersuchungen zur Theorie der Parallellinien,*[4] written in German and intended by LOBATSCHEWSKY to call the attention of mathemiaticans to his researches. Finally, in 1855, a year before his death, when he was already blind, he dictated and published in Russian and French a complete exposition of his system of geometry under the title: *Pangéométrie ou précis de géométrie fondée sur une théorie générale et rigoureuse des parallèles.*[5]

§ 40. Non-Euclidean Geometry, just as it was conceived by GAUSS and SCHWEIKART in 1816, and studied as an ab-

H. LIEBMANN, with Notes. Abhandlungen zur Geschichte der Mathematik, Bd. XIX, p. 3—50 (Leipzig, Teubner, 1904).

1 Scientific Publications of the University of Kasan (1835—38). *Geom. Works.* Vol. I: p. 219—486. German translation by F. ENGEL, p. 67—235 of his work referred to on p. 84. English translation of the Introduction by G. B. HALSTED, (Austin, Texas, 1897).

2 Scientific Publications of the University of Kasan. (1836). *Geom. Works,* Vol. I, p. 121—218. German translation by H. LIEBMANN; loc. cit: p. 51—130.

3 CRELLE's Journal, Bd. XVII, p. 295—320. (1837). *Geom. Works,* Vol. II, p. 581—613.

4 Berlin (1840). *Geom. Works,* Vol. II, p. 553—578. French translation by J. HOÜEL in Mém. de Bourdeaux, T. IV. (1866), and also in *Recherches géométriques sur la théorie des parallèles* (Paris, Hermann, 1900). English translation by G. B. HALSTED, (Austin, Texas, 1891). Facsimile reprint (Berlin, Mayer and Müller, 1887).

5 *Collection of Memoirs by Professors of the Royal University of Kasan on the 50th anniversary of its foundation.* Vol. I, p. 279—340. (1856). Also in *Geom. Works,* Vol. II, p. 617—680. In Russian, in Scientific Publications of the University of Kasan, (1855). Italian translation, by G. BATTAGLINI, in Giornale di Mat. T. V. p. 273—336, (1867). German translation, by H. LIEBMANN, Ostwald's Klassiker der exakten Wissenschaften, Nr. 130 (Leipzig, 1902).

stract system by TAURINUS in 1826, became in 1829—30
a recognized part of the general scientific inheritance.

To describe, as shortly as possible, the method followed
by LOBATSCHEWSKY in the construction of the *Imaginary Geometry* or *Pangeometry,* let us glance at his *Geometrische Untersuchungen zur Theorie der Parallellinien* of 1840.

In this work LOBATSCHEWSKY states, first of all, a group
of theorems independent of the theory of parallels. Then he
considers a pencil with vertex
A, and a straight line *BC,* in
the plane of the pencil, but
not belonging to it. Let *AD*
be the line of the pencil which
is perpendicular to *BC,* and
AE that perpendicular to
AD. In the Euclidean system

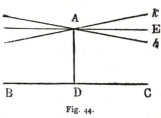

Fig. 44.

this latter line is the *only* line which does not intersect *BC.*
In the geometry of LOBATSCHEWSKY *there are other lines of the
pencil through A which do not intersect BC.* The *non-intersecting* lines are separated from the *intersecting* lines by the
two lines *h, k* (see Fig. 44), which in their turn do not meet
BC. [cf. SACCHERI, p. 42.] These lines, which the author calls
parallels, have each a definite *direction of parallelism.* The
line *h,* of the figure, is the parallel to the right: *k,* to the left.
The angle which the perpendicular *AD* makes with one of
the parallels is the *angle of parallelism* for the length *AD.*
LOBATSCHEWSKY uses the symbol Π (*a*) to denote the angle
of parallelism corresponding to the length *a.* In the ordinary
geometry, we have Π (*a*)=90° always. In the geometry of
LOBATSCHEWSKY, it is a definite function of *a,* tending to
90° as *a* tends to zero, and to zero as *a* increases without
limit.

From the definition of parallels the author then deduces
their principal properties:

That if *AE* is the parallel to *BC* for the point *A*, it is the parallel to *BC* in that direction for every point on *AE* [permanency];

That if *AE* is parallel to *BC*, then *BC* is parallel to *AE* [reciprocity] :

That if the lines (2) and (3) are parallel to (1), then (2) and (3) are parallel to each other [transitivity] [cf. GAUSS, p. 72]; and that

If *AE* and *BC* are parallel, *AE* is *asymptotic* to *BC*.

Finally, the discussion of these questions is preceded by the theorems on the sum of the angles of a triangle, the same theorems as those already given by LEGENDRE, and still earlier by SACCHERI. There can be little doubt that Lo-BATSCHEWSKY was familiar with the work of LEGENDRE.[1]

But the most important part of the *Imaginary Geometry* is the construction of the formulæ of trigonometry.

To obtain these, the author introduces two new figures: the Horocycle [circle of infinite radius, cf. GAUSS, p. 74], and the Horosphere[2] [the sphere of infinite radius], which in the ordinary geometry are the straight line and plane, respectively. Now on the Horosphere, which is made up of ∞^2 Horocycles, there exists a geometry analogous to the ordinary geometry, in which Horocycles take the place of straight lines. Thus LOBATSCHEWSKY obtains this first remarkable result:

The Euclidean Geometry [cf. WACHTER, p. 63], *and, in particular, the ordinary plane trigonometry, hold upon the Horosphere.*

---

[1] Cf. LOBATSCHEWSKY's criticism of LEGENDRE's attempt to obtain a proof of Euclid's Postulate in his *New Principles of Geometry* (ENGEL's translation, p. 68).

[2] [LOBATSCHEWSKY uses the terms *Grenzkreis, Grenzkugel* in his German work: *courbe-limite, horicycle, horisphère, surface-limite* in his French work.]

This remarkable property and another relating to *Co-axal Horocycles* [concentric circles with infinite radius] are employed by LOBATSCHEWSKY in deducing the formulæ of the new Plane and Spherical Trigonometries[1]. The formulæ of spherical trigonometry in the new system are found to be exactly the same as those of ordinary spherical trigonometry, when the elements of the triangle are measured in right-angles.

§ 41. It is well to note the form in which LOBATSCHEWSKY expresses these results. In the plane triangle ABC, let the sides be denoted by a, b, c, the angles by A, B, C; and let $\Pi(a), \Pi(b), \Pi(c)$ be the angles of parallelism corresponding to the sides a, b, c. Then LOBATSCHEWSKY's fundamental formula is

$$(4) \qquad \cos A \cos \Pi(b) \cos \Pi(c) + \frac{\sin \Pi(b) \sin \Pi(c)}{\sin \Pi(a)} = 1.$$

It is easy to see that this formula and that of TAURINUS [(1), p. 79] can be transformed into each other.

To pass from that of TAURINUS to that of LOBATSCHEWSKY, we make use of (3) of p. 80, observing that the angle B, which appears in it, is $\Pi(a)$.

For the converse step, it is sufficient to use one of LOBATSCHEWSKY's results, namely:

$$(5) \qquad \tan \frac{\Pi(x)}{2} = a.^{-x}$$

This is the same as the equation (3) of TAURINUS, under another form.

The constant a which appears in (5) is indeterminate. It represents the constant ratio of the arcs cut off two Coaxal

---

[1] It can be proved that the formulae of Non-Euclidean Plane Trigonometry can be obtained without the introduction of the *Horosphere*. The only result required is the relation between the arcs cut off two *Horocycles* by two of their axes (cf. p. 90). Cf. H. LIEBMANN, *Elementare Ableitung der nichteuklidischen Trigonometrie.* Ber. d. kön. Sach. Ges. d. Wiss., Math. Phys. Klasse, (1907).

Horocycles by a pair of axes, when the distance between
these arcs is the unit of length.
[Fig. 45.]

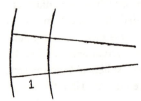

Fig. 45.

If we choose, with LOBATSCHEW-
SKY, a convenient unit, we are able
to take a equal to e, the base of
Natural Logarithms. If we wish,
on the other hand, to bring Lo-
BATSCHEWSKY's results into accord
with the *log.-spherical geometry* of TAURINUS, or the *Non-Euclidean geometry* of GAUSS, we take

$$a = e^{\frac{1}{k}}.$$

Then (5) becomes

(5′)
$$\tan \frac{\Pi(x)}{2} = e^{-\frac{x}{k}},$$

which is the same as

(6)
$$\cosh \frac{x}{k} = \frac{1}{\sin \Pi(x)}.$$

This result at once transforms LOBATSCHEWSKY's equation (4) into the equation (1) of TAURINUS.

It follows that:

*The log.-spherical geometry of Taurinus is identical with
the imaginary geometry [pangeometry] of Lobatschewsky.*

§ 42. We add the most remarkable of the results which
LOBATSCHEWSKY deduces from his formulæ:

(a) In the case of triangles whose sides are very small
[infinitesimal] we can use the ordinary trigonometrical formulæ as the formulæ of *Imaginary Trigonometry*, infinitesimals of a higher order being neglected[1].

---

[1] Conversely, the assumption that the Euclidean Geometry
holds for the infinitesimally small can be taken as the starting
point for the development of Non-Euclidean Geometry. It is one
of the most interesting discoveries from the recent examination of

(b) If for a, b, c are substituted ia, ib, ic, the formulæ of *Imaginary Trigonometry* are transformed into those of ordinary Spherical Trigonometry.[1]

(c) If we introduce a system of coordinates in two and three dimensions similar to the ordinary Cartesian coordinates, we can find the lengths of curves, the areas of surfaces, and the volumes of solids by the methods of analytical geometry.

§ 43. How was LOBATSCHEWSKY led to investigate the theory of parallels and to discover the Imaginary Geometry?

We have already remarked that BARTELS, LOBATSCHEW-SKY's teacher at Kasan, was a friend of GAUSS [p. 84]. If we now add that he and GAUSS were at Brunswick together during the two years which preceded his call to Kasan [1807], and that later he kept up a correspondence with GAUSS, the hypothesis at once presents itself that they were not without their influence upon LOBATSCHEWSKY's work.

We have also seen that before 1807 GAUSS had attempted to solve the problem of parallels, and that his efforts up till that date had not borne other fruit than the hope of overcoming the obstacles to which his researches had led him. Thus anything that BARTELS could have learned from GAUSS before 1807 would be of a negative character. As regards GAUSS's

---

GAUSS's MSS. that the *Princeps mathematicorum* had already followed this path. Cf. GAUSS, *Werke*, Bd. VIII, p. 255—264.

Both the works of FLYE St. MARIE, [*Théorie analytique sur la théorie des parallèles*, (Paris, 1871)], and of KILLING [*Die nichteuklidischen Raumformen in analytischer Behandlung*, (Leipzig, 1881)], are founded upon this principle. In addition, the formulae of trigonometry have been obtained in a simple manner by the application of the same principle, and the use of a few fundamental ideas, by M. SIMON. [Cf. M. SIMON, *Die Trigonometrie in der absoluten Geometrie*, CRELLE's Journal, Bd. 109, p. 187—198 (1892)].

[1] This result justifies the method followed by TAURINUS in the construction of his *log.-spherical geometry*.

later views, it appears quite certain that BARTELS had no news of them, so that we can be sure that LOBATSCHEWSKY created his geometry quite independently of any influence from GAUSS.[1] Other influences might be mentioned: e. g., besides LEGENDRE, the works of SACCHERI and LAMBERT, which the Russian geometer might have known, either directly or through KLÜGEL and MONTUCLA. But we can come to no definite decision upon this question[2]. In any case, the failure of the demonstrations of his predecessors, or the uselessness of his own earlier researches [1815—17], induced LOBATSCHEWSKY, as formerly GAUSS, to believe that the difficulties which had to be overcome were due to other causes than those to which until then they had been attributed. LOBATSCHEWSKY expresses this thought clearly in the *New Principles of Geometry* of 1825, where he says:

'The fruitlessness of the attempts made, since Euclid's time, for the space of 2000 years, aroused in me the suspicion that the truth, which it was desired to prove, was not contained in the data themselves; that to establish it the aid of experiment would be needed, for example, of astronomical observations, as in the case of other laws of nature. When I had finally convinced myself of the justice of my conjecture and believed that I had completely solved this difficult question, I wrote, in 1826, a memoir on this subject [*Exposition succincte des principes de la Géométrie*].'[3]

The words of LOBATSCHEWSKY afford evidence of a philosophical conception of space, opposed to that of KANT, which was then generally accepted. The Kantian doctrine considered space as a subjective intuition, a necessary presupposition of every experience. LOBATSCHEWSKY's doctrine was

---

[1] Cf. the work of F. ENGEL, quoted on p. 84. *Zweiter Teil; Lobatschefskijs Leben und Schriften.* Cap. VI, p. 373—383.

[2] Cf. SEGRE's work, quoted on p. 44.

[3] Cf. p. 67 of ENGEL's work named above.

rather allied to sensualism and the current empiricism, and compelled geometry to take its place again among the experimental sciences.[1]

§ 44. It now remains to describe the relation of LOBATSCHEWSKY's *Pangeometry* to the debated question of the Euclidean Postulate. This discussion, as we have seen, aimed at constructing the Theory of Parallels with the help of the first 28 propositions of Euclid.

So far as regards this problem, LOBATSCHEWSKY, having defined parallelism, assigns to it the distinguishing features of reciprocity and transitivity. The property of equidistance then presents itself to LOBATSCHEWSKY in its true light. Far from being indissolubly bound up with the first 28 propositions of Euclid, it contains an element entirely new.

The truth of this statement follows directly from the existence of the *Pangeometry* [a logical deductive science founded upon the said 28 propositions and on the negation of the Fifth Postulate], in which parallels *are not equidistant,* but are asymptotic. Further, we can be sure that the *Pangeometry* is a science in which the results follow logically one from the other, i. e., are free from internal contradictions. To prove this we need only consider, with LOBATSCHEWSKY, the analytical form in which it can be expressed.

This point is put by LOBATSCHEWSKY toward the end of his work in the following way:

'Now that we have shown, in what precedes, the way in which the lengths of curves, and the surfaces and volumes of solids can be calculated, we are able to assert that the Pangeometry is a complete system of geometry. A single glance

---

[1] Cf. The discourse on LOBATSCHEWSKY by A. VASILIEV, (Kasan, 1893). German translation by ENGEL in SCHLÖMILCH's Zeitschrift, Bd. XI, p. 205—244 (1895). English translation by HALSTED, (Austin, Texas, 1895).

at the equations which express the relations existing between the sides and angles of plane triangles, is sufficient to show that, setting out from them, Pangeometry becomes a branch of analysis, including and extending the analytical methods of ordinary geometry. We could begin the exposition of Pangeometry with these equations. We could then attempt to substitute for these equations others which would express the relations between the sides and angles of every plane triangle. However, in this last case, it would be necessary to show that these new equations were in accord with the fundamental notions of geometry. The standard equations, having been deduced from these fundamental notions, must necessarily be in accord with them, and all the equations which we would substitute for them, if they cannot be deduced from the equations, would lead to results contradicting these notions. Our equations are, therefore, the foundation of the most general geometry, since they do not depend on the assumption that the sum of the angles of a plane triangle is equal to two right angles.'[1]

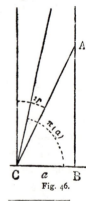

Fig. 46.

§ 45. To obtain fuller knowledge of the nature of the constant k contained implicity in LOBATSCHEWSKY's formulæ, and explicitly in those of TAURINUS, we must apply the new trigonometry to some actual case. To this end LOBATSCHEWSKY used a triangle ABC, in which the side BC (a) is equal to the radius of the earth's orbit, and A is a fixed star, whose direction is perpendicular to BC (Fig. 46). Denote by $2p$ the maximum parallax of the star A. Then we have

---

[1] Cf. the Italian translation of the *Pangéométrie*, Giornale di Mat., T. V. p. 334; or p. 75 of the German translation referred to on p. 86.

$$\Pi(a) > \measuredangle\, BAC = \frac{\pi}{2} - 2p.$$

Therefore

$$\tan \tfrac{1}{2}\, \Pi(a) > \tan\left(\frac{\pi}{4} - p\right) = \frac{1 - \tan p}{1 + \tan p}.$$

But $$\tan \tfrac{1}{2}\, \Pi(a) = e^{-\frac{a}{k}} \quad [\text{cf. p. 90}].$$

Therefore $$e^{\frac{a}{k}} < \frac{1 + \tan p}{1 - \tan p}.$$

But on the hypothesis $p < \dfrac{\pi}{4}$, we have

$$\frac{a}{k} < log\, \frac{1 + \tan p}{1 - \tan p} = 2\,(\tan p + \tfrac{1}{3}\tan^3 p + \tfrac{1}{5}\tan^5 p + \ldots).$$

Also, $\tan 2p = \dfrac{2\tan p}{1 - \tan^2 p}$

$$= 2\,(\tan p + \tan^3 p + \tan^5 p + \ldots).$$

Therefore we have

$$\frac{a}{k} < \tan 2p.$$

Take now, with LOBATSCHEWSKY, the parallax of *Sirius* as $1'',24$.

From the value of $\tan 2p$, we have

$$\frac{a}{k} < 0,0000006012.$$

This result does not allow us to assign a value to k, but it tells us that it is very great compared with the diameter of the earth's orbit. We could repeat the calculation for much smaller parallaxes, for example $0'',1$, and we would find k to be greater than a million times the diameter of the earth's orbit.

Thus, if the Euclidean Geometry and the Fifth Postulate are to hold in actual space, k must be infinitely great. That is to say, there must be stars whose parallaxes are indefinitely small.

However it is evident that we can never state whether this is the case or not, since astronomical observations will always be true only within certain limits. Yet, knowing the enormous size of k in comparison with measurable lengths, we must, with LOBATSCHEWSKY, admit that the Euclidean hypothesis is valid for all practical purposes.

We would reach the same conclusion if we regarded the question from the standpoint of the sum of the angles of a triangle. The results of astronomical observations show that the defect of a triangle, whose sides approach the distance of the earth from the sun, cannot be more than $0'',0003$. Let us now consider, instead of an astronomical triangle, one drawn on the Earth's surface, the angles of which can be directly measured. In consequence of the fundamental theorem that the area of a triangle is proportional to its defect, the possible defect would fall within the limits of experimental error. Thus we can regard the defect as zero in experimental work, and Euclid's Postulate will hold in the domain of experience.[1]

Johann Bolyai [1802—1860].

§ 46. J. BOLYAI a Hungarian officer in the Austrian army, and son of WOLFGANG BOLYAI, shares with LOBAT-SCHEWSKY the honour of the discovery of Non-Euclidean geometry. From boyhood he showed a remarkable aptitude for mathematics, in which his father himself instructed him. The teaching of WOLFGANG quickly drew JOHANN's attention to Axiom XI. To its demonstration he set himself, in spite of the advice of his father, who sought to dissuade him from the attempt. In this way the theory of parallels formed the favourite occupation of the young mathematician, during his course [1817—22] in the Royal College for Engineers at Vienna.

---

[1] For the contents of this section, cf. LOBATSCHEWSKY, *On the Principles of Geometry*. See p. 22—24 of ENGEL's work named on p. 84. Also ENGEL's remarks on p. 248—252 of the same work.

At this time JOHANN was an intimate friend of CARL SZÁSZ [1798—1853] and the seeds of some of the ideas, which led BOLYAI to create the *Absolute Science of Space*, were sown in the conversations of the two eager students.

It appears that to SZÁSZ is due the distinct idea of considering the parallel through *B* to the line *AM* as the limiting position of a secant *BC* turning in a definite direction about *B*; that is, the idea of considering *BC* as parallel to *AM*, when *BC*, in the language of SZÁSZ, *detaches itself* (springs away) from *AM* (Fig. 47). To this parallel BOLYAI gave the name of *asymptotic parallel or asymptote*. [cf. SACCHERI]. From the conversations of the two friends were also derived the conception of the *line equidistant from a straight line*, and the other most important idea of the *Paracycle* (*limiting curve* or *horocycle* of LOBATSCHEWSKY). Further they recognised that the proof of Axiom XI would be obtained if it could be shown that the *Paracycle* is a straight line.

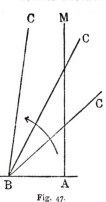

Fig. 47.

When SZÁSZ left Vienna in the beginning of 1821 to undertake the teaching of Law at the College of Nagy-Enyed (Hungary), JOHANN remained to carry on his speculations alone. Up till 1820 he was filled with the idea of finding a proof of Axiom XI, following a path similar to that of SACCHERI and LAMBERT. Indeed his correspondence with his father shows that he thought he had been successful in his aim.

The recognition of the mistakes he had made was the cause of JOHANN's decisive step towards his future discoveries, since he realised 'that one must do no violence to nature, nor model it in conformity to any blindly formed chimæra;

that, on the other hand, one must reguard nature reasonably and naturally, as one would the truth, and be contented only with a representation of it which errs to the smallest possible extent.'

JOHANN BOLYAI, then, set himself to construct an *absolute theory* of space, following the classical methods of the Greeks: that is, keeping the deductive method, but without deciding *a priori* on the truth or error of the FifthPostulate.

§ 47. As early as 1823 BOLYAI had grasped the real nature of his problem. His later additions only concerned the material and its formal expression. At that date he had discovered the formula:

$$e^{-\frac{a}{k}} = \tan \frac{\Pi(a)}{2},$$

connecting the angle of parallelism with the line to which it corresponds [cf. LOBATSCHEWSKY, p. 89]. This equation is the key to all Non-Euclidean Trigonometry. To illustrate the discoveries which JOHANN made in this period, we quote the following extract from a letter which he wrote from Temesvár to his father, on Nov. 3, 1823: 'I have now resolved to publish a work on the theory of parallels, as soon as I shall have put the material in order, and my circumstances allow it. I have not yet completed this work, but the road which I have followed has made it almost certain that the goal will be attained, if that is at all possible: the goal is not yet reached, but I have made such wonderful discoveries that I have been almost overwhelmed by them, and it would be the cause of continual regret if they were lost. When you will see them, you too will recognize it. In the meantime I can say only this: *I have created a new universe from nothing.* All that I have sent íyou till now is but a house of cards compared to the tower. I am as fully persuaded that it will bring me honour, as if I had already completed the discovery.'

WOLFGANG expressed the wish at once to add his son's theory to the *Tentamen* since 'if you have really succeeded in the question, it is right that no time be lost in making it public, for two reasons: first, because ideas pass easily from one to another, who can anticipate its publication; and secondly, there is some truth in this, that many things have an epoch, in which they are found at the same time in several places, just as the violets appear on every side in spring. Also every scientific struggle is just a serious war, in which I cannot say when peace will arrive. Thus we ought to conquer when we are able, since the advantage is always to the first comer.'

Little did WOLFGANG BOLYÀI think that his presentiment would correspond to an actual fact (that is, to the simultaneous discovery of Non-Euclidean Geometry by the work of GAUSS, TAURINUS, and LOBATSCHEWSKY).

In 1825 JOHANN sent an abstract of his work, among others, to his father and to J. WALTER VON ECKWEHR [1789—1857], his old Professor at the Military School. Also in 1829 he sent his manuscript to his father. WOLFGANG was not completely satisfied with it, chiefly because he could not see why an indeterminate constant should enter into JOHANN's formulæ. None the less father and son were agreed in publishing the new theory of space as an appendix to the first volume of the *Tentamen*:—

The title of JOHANN BOLYAI's work is as follows.

Appendix scientiam spatii absolute veram exhibens: a veritate aut falsitate Axiomatis XI. Euclidei, a priori haud unquam decidenda, independentem: adjecta ad casum falsitatis quadratura circuli geometrica.[1]

---

[1] A reprint—*Édition de Luxe*—was issued by the Hungarian Academy of Sciences, on the occasion of the first centenary of the birth of the author (Budapest, 1902). See also the English

The Appendix was sent for the first time [June, 1831] to GAUSS, but did not reach its destination; and a second time, in January, 1832. Seven weeks later (March 6, 1832), GAUSS replied to WOLFGANG thus:

"If I commenced by saying that I *am unable to praise this work* (by JOHANN), you would certainly be surprised for a moment. But I cannot say otherwise. To praise it, would be to praise myself. Indeed the whole contents of the work, the path taken by your son, the results to which he is led, coincide almost entirely with my meditations, which have occupied my mind partly for the last thirty or thirty-five years. So I remained quite stupefied. So far as my own work is concerned, of which up till now I have put little on paper, my intention was not to let it be published during my lifetime. Indeed the majority of people have not clear ideas upon the questions of which we are speaking, and I have found very few people who could regard with any special interest what I communicated to them on this subject. To be able to take such an interest it is first of all necessary to have devoted careful thought to the real nature of what is wanted and upon this matter almost all are most uncertain. On the other hand it was my idea to write down all this later so that at least it should not perish with me. It is therefore a pleasant surprise for me that I am spared this trouble, and I am very glad that it is just the son of my old friend, who takes the precedence of me in such a remarkable manner."

WOLFGANG communicated this letter to his son, adding: "GAUSS's answer with regard to your work is very satis-

---

translation by HALSTED, *The Science Absolute of Space,* (Austin, Texas, 1896). An Italian translation by G. B. BATTAGLINI appeared in the Giornale di Mat., T. VI, p. 97—115 (1868). Also a French translation by HOÜEL, in Mém. de la Soc. des Sc. de Bordeaux, T. V. p. 189—248 (1867). Cf. also FRISCHAUF, *Absolute Geometrie nach Johann Bolyai,* (Leipzig, Teubner, 1872).

factory and redounds to the honour of our country and of our nation."

Altogether different was the effect GAUSS's letter produced on JOHANN. He was both unable and unwilling to convince himself that others, earlier than and independent of him, had arrived at the *Non-Euclidean Geometry*. Further he suspected that his father had communicated his discoveries to GAUSS before sending him the *Appendix* and that the latter wished to claim for himself the priority of the discovery. And although later he had to let himself be convinced that such a suspicion was unfounded, JOHANN always regarded the "Prince of Geometers" with an unjustifiable aversion.[1]

§ 48. We now give a short description of the most important results contained in JOHANN BOLYAI's work:

a) The definition of parallels and their properties independent of the Euclidean postulate.

b) The circle and sphere of infinite radius. The geometry on the sphere of infinite radius is identical with ordinary plane geometry.

c) Spherical Trigonometry is independent of Euclid's Postulate. Direct demonstration of the formulæ.

d) Plane Trigonometry in Non-Euclidean Geometry. Applications to the calculation of areas and volumes.

e) Problems which can be solved by elementary methods. Squaring the circle, on the hypothesis that the Fifth Postulate is false.

While LOBATSCHEWSKY has given the Imaginary Geometry a fuller development especially on its analytical side,

---

[1] For the contents of this and the preceding article see STÄCKEL, *Die Entdeckung der nichteuklidischen Geometrie durch Johann Bolyai.* Math. u. Naturw. Berichte aus Ungarn. Bd. XVII, [1901].

Also STÄCKEL u. ENGEL. *Gauss, die beiden Bolyai und die nichteuklidische Geometrie.* Math. Ann. Bd. XLIX, p. 149—167 [1897]. Bull. Sc. Math. (2) T. XXI, pp. 206—228 [1897].

BOLYAI entered more fully into the question of the dependence or independence of the theorems of geometry upon Euclid's Postulate. Also while LOBATSCHEWKY chiefly sought to construct a system of geometry on the negation of the said postulate, JOHANN BOLYAI brought to light the propositions and constructions in ordinary geometry which are independent of it. Such propositions, which he calls *absolutely true*, pertain to the *absolute science* of space. We could find the propositions of this science by comparing Euclid's Geometry with that of LOBATSCHEWSKY. Whatever they have in common, e. g. the formulæ of Spherical Trigonometry, pertains to the Absolute Geometry. JOHANN BOLYAI, however, does not follow this path. He shows directly, that is independently of the Euclidean Postulate, that his propositions are absolutely true.

§ 49.   One of BOLYAI's absolute theorems, remarkable for its simplicity and neatness, is the following:

The sines of the angles of a rectilinear triangle are to one another as the circumferences of the circles whose radii are equal to the opposite sides.

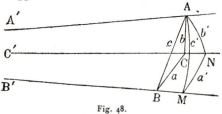

Fig. 48.

Let *ABC* be a triangle in which *C* is a right angle, and *BB′* the perpendicular through *B* to the plane of the triangle.

Draw the parallels through *A* and *C* to *BB′* in the same sense.

Then let the Horosphere be drawn through *A* (eventually the plane) cutting the lines *AA′*, *BB′* and *CC′*, respectively, in the points *A, M,* and *N*.

If we denote by a', b', c' the sides of the rectangular triangle AMN on the Horosphere, it follows from what has been said above [cf. § 48 (b)] that

$$\sin AMN = \frac{b'}{c'}.$$

But two arcs of Horocycles on the Horosphere are proportional to the circumferences of the circles which have these arcs for their (horocyclic) radii.

If we denote by *circumf.* x' the circumference of the circle whose (horocyclic) radius is x', we can write:

$$\sin AMN = \frac{\text{circumf. } b'}{\text{circumf. } c'}.$$

On the other hand, the circle traced on the Horosphere with horocyclic radius of length x', can be regarded as the circumference of an odinary circle whose radius (rectilinear) is half of the chord of the arc $2\,x'$ of the Horocycle.

Denoting by $\bigcirc x$ the circumference of the circle whose (rectilinear) radius is x, and observing that the angles ABC and AMN are equal, the preceding equation taken from

$$\sin ABC = \frac{\bigcirc b}{\bigcirc c}.$$

From the property of the right angled triangle ABC expressed by this equation, we can deduce BOLYAI's theorem enunciated above, just as from the Euclidean equation

$$\sin ABC = \frac{b}{c}$$

we can deduce that the sines of the angles of a triangle are proportional to the opposite sides. [*Appendix* § 25.]

BOLYAI's Theorem may be put shortly thus:

(1) $\qquad \bigcirc a : \bigcirc b : \bigcirc c = \sin A : \sin B : \sin C.$

If we wish to discuss the geometrical systems separately we will have

(i) In the case of the Euclidean Hypothesis,

$$\bigcirc x = 2\,\pi x.$$

Thus, substituting in (1), we have

(1′) $a : b : c : = \sin A : \sin B : \sin C.$

(ii)   In the case of the Non-Euclidean Hypothesis,

$$\bigcirc x = \pi k \left(e^{\frac{x}{k}} - e^{-\frac{x}{k}} \right) = 2 \pi k \sinh \frac{x}{k}.$$

Then substituting in (1) we have

(1″) $\sinh \frac{a}{k} : \sinh \frac{b}{k} : \sinh \frac{c}{k} = \sin A : \sin B : \sin C.$

This last relation may be called the *Sine Theorem of the Bolyai-Lobatschewsky Geometry.*

From the formula (1) Bolyai deduces, in much the same way as the usual relations are obtained from (1′), *the proportionality of the sines of the angles and the opposite sides in a spherical triangle.* From this it follows that Spherical Trigonometry is independent of the Euclidean Postulate [*Appendix* § 26].

This fact makes the importance of Bolyai's Theorem still clearer.

§ 50. The following construction for a parallel through the point D to the straight line AN belongs also to the Absolute Geometry [*Appendix* § 34].

Draw the perpendiculars DB and AE to AN [Fig. 49].

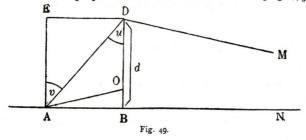

Fig. 49.

Also the perpendicular DE to the line AE. The angle EDB of the quadrilateral $ABDE$, in which three angles are right angles, is a right angle or an acute angle, according as ED is equal to or greater than AB.

With centre A describe a circle whose radius is equal to ED.

It will intersect DB at a point O, coincident with B or situated between B and D.

The angle which the line AO makes with DB is the angle of parallelism corresponding to the segment BD.[1] [*Appendix* § 27.]

Therefore a parallel to AN through D can be constructed by drawing the line DM so that $\measuredangle BDM$ is equal to $\measuredangle AOB$.[2]

---

[1] We give a sketch of BOLYAI's proof of this theorem: The circumferences of the circles with radii AB and ED, traced out by the points B and D in their rotation about the line AE, can be considered as belonging, the first to the plane through A perpendicular to the axis AE, the second to an Equidistant Surface for this plane. The constant distance between the surface and the plane is the segment $BD = d$. The ratio between these two circumferences is thus a function of d only. Using BOLYAI's Theorem, § 49, and applying it to the two rightangled triangles ADE and ADB, this ratio can be expressed as

$$\bigcirc AB : \bigcirc ED = \sin u : \sin v.$$

From this it is clear that the ratio $\sin u : \sin v$ does not vary if the line AE changes its position, remaining always perpendicular to AB, while d remains fixed. In particular, if the foot of AE tends to infinity along AN, u tends to $\Pi(d)$ and v to a right angle. Consequently,

$$\bigcirc AB : \bigcirc ED = \sin \Pi(d) : 1.$$

On the other hand in the right-angled triangle AOB, we have the equation

$$\bigcirc AB : \bigcirc AO = \sin AOB : 1.$$

This, with the preceding equation, is sufficient to establish the equality of the angles $\Pi(d)$ und AOB.

[2] Cf. *Appendix III* to this volume.

§ 51. The most interesting of the Non-Euclidean constructions given by BOLYAI is that for the *squaring of the
circle*. Without keeping strictly to BOLYAI's method, we shall
explain the principal features of his construction.

But we first insert the converse of the construction of
§ 50, which is necessary for our purpose.

*On the Non-Euclidean Hypothesis to draw the segment
which corresponds to a given (acute) angle of parallelism.*

Assuming that the theorem, that the three perpendiculars
from the angular points of a triangle on the opposite sides
intersect *eventually*, is also true in the Geometry of BOLYAI-
LOBATSCHEWSKY, on the line AB bounding the acute angle
BAA' take a point B, such that the parallel BB' to AA'
through B makes an acute angle (ABB') with AB. [Fig. 50.]

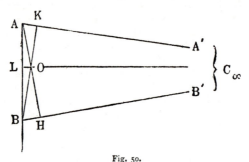

Fig. 50.

The two rays AA', BB', and the line AB may be
regarded as the three sides of a triangle of which one angular
point is C_∞, common to the two parallels AA', BB'. Then
the perpendiculars from A, B, to the opposite sides, meet in
he point O inside the triangle, and the perpendicular from
C_∞ to AB also passes through O.

Thus, if the perpendicular OL is drawn from O to AB,
the segment AL will have been found which corresponds to
the angle of parallelism BAA'.

As a particular case the angle BAA' could be 45°. Then AL would be *Schweikart's Constant* [cf. p. 76].

We note that the problem which we have just solved could be enunciated thus:

To draw a line which shall be parallel to one of the lines bounding an acute angle and perpendicular to the other.[1]

§ 52. We now show how the preceding result is used *to construct a square equal in area to the maximum triangle.*

The area of a triangle being

$$k^2 (\pi - \measuredangle A - \measuredangle B - \measuredangle C),$$

the maximum triangle, i. e. that for which the three angular points are at infinity, will have for area

$$\Delta = k^2 \pi.$$

To find the angle ω of a square whose area is $k^2\pi$, we need only remember (LAMBERT, p. 46) that the area of a polygon, as well as of a triangle, is proportional to its defect. Thus we have the equation

$$k^2 \pi = k^2 (2\pi - 4\omega),$$

from which it follows that

$$\omega = \frac{1}{4} \pi = 45°.$$

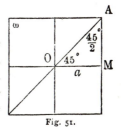

Fig. 51.

Assuming this, let us consider the right-angled triangle OAM (Fig. 51), which is the eighth part of the required square. Putting $OM = a$, and applying the formula (2) of p. 80 we obtain

$$\cosh \frac{a}{k} = \frac{\cos 22° 30'}{\sin 45°},$$

$$\text{or } \cosh \frac{a}{k} = \frac{\sin 67° 30'}{\sin 45°}.$$

---

[1] BOLYAI's solution [*Appendix.*, § 35] is, however, more complicated.

If we now draw, as in § 51, the two segments b', c', which correspond to the angles $67°\ 30'$ and $45°$, and if we remember that [cf. p. 90 (6)]

$$\cosh \frac{x}{k} = \frac{1}{\sin \Pi (x)},$$

the following relation must hold between a, b' and c',

$$\cosh \frac{a}{k} \cosh \frac{b'}{k} = \cosh \frac{c'}{k}.$$

Finally if we take b' as side, and c' as hypotenuse of a right-angled triangle, the other side of this triangle, by formula (1) of p. 79, is determined by the equation

$$\cosh \frac{a'}{k} \cosh \frac{b'}{k} = \cosh \frac{c'}{k}.$$

Then comparing these two questions, we obtain

$$a' = a.$$

Constructing a in this way, we can immediately find the square whose area is equal to that of the maximum triangle.

§ 53. To construct a circle whose area shall be equal to that of this square, that is, to the area of the maximum triangle, we must transform the expression for the area of a circle of radius r

$$2 \pi k^2 \left(\cosh \frac{r}{k} - 1 \right),$$

given on p. 81, by the introduction of the angle of parallelism $\Pi\left(\dfrac{r}{2}\right)$, corresponding to half the radius.

Then we have [1] for the area of this circle

$$\frac{4 \pi k^2}{\tan^2 \Pi \left(\dfrac{r}{2} \right)}$$

On the other hand if the two parallels AA' and BB' are drawn from the ends of the segment AB, making equal angles with AB, we have

---

[1] Using the result $\tan \dfrac{\Pi (x)}{2} = e .^{- x/k}$

$$\measuredangle\,A'AB = \measuredangle\,B'BA = \Pi\left(\frac{r}{2}\right),$$

where $AB = r$ [Fig. 52].

Now draw AC, perpendicular to BB', and AD perpendicular to AC; also put

$$\measuredangle\,CAB = \alpha,\ \measuredangle\,DAA' = z.$$

Then we have

$$\tan z = \cot\left(\Pi\left(\frac{r}{2}\right) - \alpha\right) = \frac{\cot\Pi\left(\frac{r}{2}\right)\cot\alpha + 1}{\cot\alpha - \cot\Pi\left(\frac{r}{2}\right)}.$$

It is easy to eliminate α from this last result by means of the trigonometrical formulae for the triangle ABC and so obtain

$$\tan z = \frac{2}{\tan\Pi\left(\frac{r}{2}\right)}$$

Substituting this in the expression found for the area of the circle, we obtain for that area

$$\pi\,k^2\,\tan^2 z.$$

This formula, proved in another way by Bolyai [*Appendix* § 43], allows us to associate a definite angle z with every circle. If z were equal to $45°$, then we would have

$$\pi\,k^2$$

for the area of the corresponding circle.

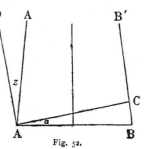

Fig. 52.

---

[1] Indeed, in the rightangled triangle ABC, we have $\cot\Pi\left(\frac{r}{2}\right)$ $\cot\alpha = \cosh\dfrac{r}{k}$. From this, since $\cosh\dfrac{r}{k} = 2\sinh^2\dfrac{r}{2k} + 1 = 2\cot^2\Pi\left(\frac{r}{2}\right) + 1$, we deduce, first, that

That is: *the area of the circle, for which the angle z is 45°, is equal to the area of the maximum triangle*, and thus to that of the square of § 52.

If $z = \sphericalangle A'AD$ (Fig. 51) is given, we can find r by the following construction:

    (i)   Draw the line AC perpendicular to AD.

    (ii)   Draw BB' parallel to AA' and perpendicular to AC (§ 51).

    (iii)   Draw the bisector of the strip between AA' and BB'.

[By the theorem on the concurrency of the bisectors of the angles of a triangle with an *infinite* vertex.]

    (iv)   Draw the perpendicular AB to this bisector. The segment AB bounded by AA' and BB' is the required radius r.

§ 54. The problem of constructing a polygon equal to a circle of area $\pi k^2 \tan^2 z$ is, as BOLYAI remarked, closely allied with the numerical value of tan z. It is resolvable for every integral value of $\tan^2 z$, and for every fractional value, provided that the denominator of the fraction, reduced to its lowest terms, is included in the form assigned by GAUSS for the construction of regular polygons [*Appendix* § 43].

The possibility of constructing a square equal to a circle leads JOHANN to the conclusion *"habeturque aut Axioma XI Euclidis verum, aut quadratura circuli geometrica;*

$$\cot \Pi\left(\frac{r}{2}\right) \cot \alpha = 2 \cot^2 \Pi\left(\frac{r}{2}\right) + 1,$$

and next that

$$\cot \alpha - \cot \Pi\left(\frac{r}{2}\right) = \left(1 + \tan^2 \Pi\left(\frac{r}{2}\right)\right) \cot \Pi\left(\frac{r}{2}\right).$$

These equations allow the expression for tan z to be written down in the required form.

etsi hucusque indecisum manserit, quodnam ex his duobus revera locum habeat."

This dilemma seemed to him at that time [1831] impossible of solution, since he closed his work with these words: "Superesset denique (ut res omni numero absolvatur), impossibilitatem (absque suppositione aliqua) decidenda, num Σ (the Euclidean system) aut aliquod (et quodam) S (the Non-Euclidean system) sit, demonstrare: quod tamen occasioni magis idoneae reservatur."

JOHANN, however, never published any demonstration of this kind.

§ 55. After 1831 BOLYAI continued his labours at his geometry, and in particular at the following problems:

1. The connection between Spherical Trigonometry and Non-Euclidean Trigonometry.

2. Can one prove rigorously that EUCLID's Axiom is not a consequence of what precedes it?

3. The volume of a tetrahedron in Non-Euclidean geometry.

As regards the first of these problems, beyond establishing the analytical relation connecting the two trigonometries [cf. LOBATSCHEWSKY, p. 90], BOLYAI recognized that in the Non-Euclidean hypothesis there exist three classes of *Uniform* Surfaces[1] on which the Non-Euclidean trigonometry, the ordinary trigonometry, and spherical trigonometry respectively hold. To the first class belong planes and *hyperspheres* [surfaces equidistant from a plane]; to the second, the *paraspheres* [LOBATSCHEWSKY's Horospheres]; to the third, spheres. The paraspheres are the limiting case when we pass from the hyperspherical surfaces to the spherical. This passage is shown analytically by making a

---

[1] BOLYAI seems to indicate by this name the surfaces which behave as planes, with respect to displacement upon themselves.

certain parameter, which appears in the formulæ, vary continuously from the real domain to the purely imaginary through infinity [cf. TAURINUS, p. 82].

As to the second problem, that regarding the impossibility of demonstrating *Axiom XI,* BOLYAI neither succeeded in solving it, nor in forming any definite opinion upon it. For some time he believed that we could not, in any way, decide which was true, the Euclidean hypothesis or the Non-Euclidean. Like LOBATSCHEWSKY, he relied upon the analytical possibility of the new trigonometry. Then we find JOHANN returning again to the old ideas, and attempting a new demonstration of *Axiom XI.* In this attempt he applies the Non-Euclidean formulæ to a system of five coplanar points. There must necessarily be some relation between the distance of these points. Owing to a mistake in his calculations JOHANN did not find this relation, and for some time he believed that he had proved, in this way, the falsehood of the Non-Euclidean hypothesis and the absolute truth of *Axiom XI.*[1]

However he discovered his mistake later, but he did not carry out further investigations in this direction, as the method, when applied to six or more points, would have involved too complicated calculations.

The third of the problems mentioned above, that regarding the tetrahedron, is of a purely geometrical nature. BOLYAI's solutions have been recently discovered and pub-

---

[1] The title of the paper which contains JOHANN's demonstration is as follows: *"Beweis des bis nun auf der Erde immer noch zweifelhaft gewesenen, weltberühmten und, als der gesammten Raum- und Bewegungslehre zu Grunae dienend, auch in der That allerhöchstwichtigsten 11. Euclid'schen Axioms* von *J. Bolyai von Bolya, k. k. Genie-Stabshauptmann in Pension.* Cf. STÄCKEL's paper: *Untersuchungen aus der Absoluten Geometrie aus Johann Bolyais Nachlaß.* Math. u. Naturw. Berichte aus Ungarn. Bd. XVIII, p. 280—307 (1902). We are indebted to this paper for this section § 55.

lished by STÄCKEL [cf. p. 112 note 1]. LOBATSCHEWSKY had been often occupied with the same problem from 1829[1], and GAUSS proposed it to JOHANN in his letter quoted on p. 100.

Finally we add that J. BOLYAI heard of LOBATSCHEWSKY's *Geometrische Untersuchungen* in 1848: that he made them the object of critical study[2]: and that he set himself to compose an important work on the reform of the Principles of Mathematics with the hope of prevailing over the Russian. He had planned this work at the time of the publication of the *Appendix*, but he never succeeded in bringing it to a conclusion.[3]

The Absolute Trigonometry.

§ 56. Although the formulæ of Non-Euclidean trigonometry contain the ordinary relations between the sides and angles of a triangle as a limiting case [cf. p. 80], yet they do not form a part of what JOHANN BOLYAI called *Absolute Geometry*. Indeed the formulæ do not apply at once to the two classes of geometry, and they were deduced on the supposition of the validity of the *Hypothesis of the Acute Angle*. Equations directly applicable both to the Euclidean case and to the Non-Euclidean case were met by us in § 49 and they make up BOLYAI's Theorem. They are three in number, only two of them being independent. Thus they furnish a first set of formulæ of *Absolute Trigonometry*.

---

[1] Cf. p. 53 et seq., of the work quoted on p. 84. Also LIEBMANN's translation, referred to in Note 2, p. 85.

[2] Cf. P. STÄCKEL und J. KÜRSCHÁK: *Johann Bolyais Bemerkungen über N. Lobatschefskijs Geometrische Untersuchungen zur Theorie der Parallellinien*, Math. u. Naturw. Berichte aus Ungarn, Bd. XVIII, p. 250—279 (1902).

[3] Cf. P. STÄCKEL: *Johann Bolyais Raumlehre*, Math. u. Naturw. Berichte aus Ungarn, Bd. XIX (1903).

Other formulæ of Absolute Trigonometry were given in 1870 by the Belgian geometer, DE TILLY, in his *Études de Mécanique Abstraite.*[1]

The formulæ given by DE TILLY refer to rectilinear triangles, and were deduced by means of kinematical considerations, requiring only those properties of a bounded region of a plane area, which are independent of the value of the sum of the angles of a triangle.

In addition to the function $\bigcirc x$, which we have already met in BOLYAI's formulæ, those of DE TILLY contain another function Ex defined in the following way:

Let r be a straight line and l the *equidistant curve,* distant x from r. Since the arcs of l are proportional to their projections on r, it is clear that the ratio between a (rectified) arc of l and its projection does not depend on the length of the arc, but only on its distance x from r. DE TILLY's function Ex is the function which expresses this ratio.

On this understanding, the Formulæ of Absolute Trigonometry for the right-angled triangle ABC are as follows:

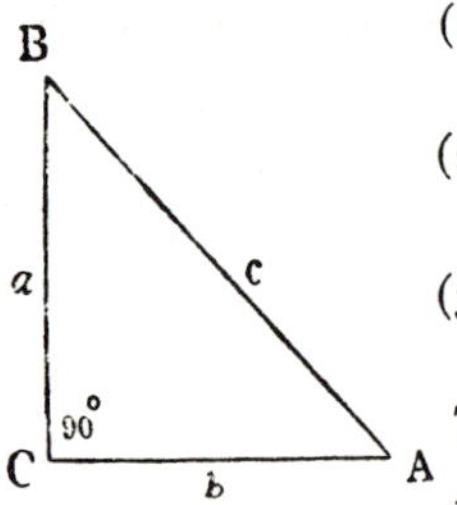

Fig. 53.

(1)
$$\begin{cases} \bigcirc a = \bigcirc c \sin A \\ \bigcirc b = \bigcirc c \sin B \end{cases}$$

(2)
$$\begin{cases} \cos A = Ea.\ \sin B \\ \cos B = Eb.\ \sin A \end{cases}$$

(3)
$$Ec = Ea.\ Eb.$$

The set (1) is equivalent to BOLYAI's Theorem for the Right-Angled Triangle. All the formulæ of Absolute Trigonometry could be derived by suitable combination of these three sets. In particular, for the right-angled triangle, we obtain the following equation:—

---

[1] Mémoires couronnés et autres Mémoires, Acad. royale de Belgique. T. XXI (1870). Cf. also the work by the same author: *Essai sur les principes fondamentaux de la géométrie et de la Mécanique,* Mém. de la Soc. des Sc. de Bordeaux. T. III (cah. I) (1878).

$$\bigcirc^2 a\,(Ea + Eb.\,Ec) + \bigcirc^2 b.\,(Eb + Ec.\,Ea)$$
$$= \bigcirc^2 c\,(Ec + Ea.\,Eb).$$

This can be regarded as equivalent to the Theorem of Pythagoras in the Absolute Geometry.[1]

§ 57. Let us now see how we can deduce the results of the Euclidean and Non-Euclidean geometries from the equations of the preceding article.

Euclidean Case.

The *Equidistant Curve (l)* is a straight line [that is, $Ex = 1$], and the perimeters of circles are proportional to their radii.

Thus the equations (1) become

(1')
$$\begin{cases} a = c \sin A \\ b = c \sin B. \end{cases}$$

The equations (2) give

(2') $\cos A = \sin B,$ $\cos B = \sin A.$

Therefore $A + B = 90^\circ.$

Finally the equation (3) reduces to an identity.

The equations (1') and (2') include the whole of ordinary trigonometry.

Non-Euclidean Case.

Combining the equations (1) and (2) we obtain

(5)
$$\frac{\bigcirc^2 a}{E^2 a - 1} = \frac{\bigcirc^2 b}{E^2 b - 1}.$$

If we now apply the first of equations (2) to a right-angled triangle whose vertex A goes off to infinity, so that the angle A tends to zero, we shall have

$$Lt \cos A = Lt\,(Ea.\,\sin B).$$

But Ea is independent of A; also the angle B, in the limit, becomes the angle of parallelism corresponding to a, i. e. $\Pi\,(a).$

---

[1] Cf. R. Bonola, *La trigonometria assoluta secondo Giovanni Bolyai.* Rend. Istituto Lombardo (2). T. XXXVIII (1905).

Therefore we have

$$Ea = \frac{1}{\sin \Pi(a)}.$$

A similar result holds for Eb.

Substituting these in equation (5) we obtain

$$\frac{\bigcirc^2 a}{\cot^2 \Pi(a)} = \frac{\bigcirc^2 b}{\cot^2 \Pi(b)},$$

from which

$$\frac{\bigodot a}{\cot \pi(a)} = \frac{\bigcirc b}{\cot \pi(b)}.$$

This result, with the expression for Ex, allows us at once to obtain from the equations (1), (2), (3), the formulæ of the Trigonometry of BOLYAI-LOBATSCHEWSKY:

(1″) $\begin{cases} \cot \Pi(a) = \cot \Pi(c) \sin A \\ \cot \Pi(b) = \cot \Pi(c) \sin B, \end{cases}$

(2″) $\begin{cases} \sin A = \cos B \sin \Pi(b) \\ \sin B = \cos A \sin \Pi(a), \end{cases}$

(3″) $\sin \Pi(c) = \sin \Pi(a) \sin \Pi(b).$

These relations between the elements of every right-angled triangle were given in this form by LOBATSCHEWSKY.[1] If we wish to introduce direct functions of the sides, instead of the angles of parallelism $\Pi(a)$, $\Pi(b)$ and $\Pi(c)$, it is sufficient to remember [p. 90] that

$$\tan \frac{\Pi(x)}{2} = e^{-x/k}.$$

We can thus express the circular functions of $\Pi(x)$ in terms of the hyperbolic functions of x. In this way the preceding equations are replaced by the following relations:

(1‴) $\sinh \dfrac{a}{k} = \sinh \dfrac{c}{k} \sin A$

$\sinh \dfrac{b}{k} = \sinh \dfrac{c}{k} \sin B,$

---

[1] Cf. e. g., The *Geometrische Untersuchungen* of LOBATSCHEWSKY referred to on p. 86.

(2''') $$\cos A = \sin B \cosh \frac{a}{k}$$

$$\cos B = \sin A \cosh \frac{b}{k},$$

and

(3''') $$\cosh \frac{c}{k} = \cosh \frac{a}{k} \cos h \cosh \frac{b}{k}.$$

§ 58. The following remark upon Absolute Trigonometry is most important: *If we regard the elements in its formulae as elements of a spherical triangle, we obtain a system of equations which hold also for Spherical Triangles.*

This property of Absolute Trigonometry is due to the fact, already noticed on p. 114, that it was obtained by the aid of equations which hold only for a limited region of the plane. Further these do not depend on the hypothesis of the angles of a triangle, so that they are valid also on the sphere.

If it is desired to obtain the result directly, it is only necessary to note the following facts:—

(i) In Spherical Trigonometry the circumferences of circles are proportional to the sines of their (spherical) radii, so that the first formula for right-angled spherical triangles

$$\sin a = \sin c \sin A$$

is transformed at once into the first of the equations (1).

(ii) A circle of (spherical) radius $\frac{\pi}{2} - b$ can be considered as a curve equidistant from the concentric great circle, and the ratio Eb for these two circles is given by

$$\frac{\sin \left(\frac{\pi}{2} - b \right)}{\sin \frac{\pi}{2}} = \cos b.$$

Thus the formulæ for right-angled spherical triangles

$$\cos A = \sin B \cos a,$$
$$\cos c = \cos a \cos b,$$

are transformed immediately into the equations (2) and (3) by means of this result.

Thus *the formulae of Absolute Trigonometry also hold on the sphere.*

Hypotheses equivalent to Euclid's Postulate.

§ 59. Before leaving the elementary part of the subject, it seems right to call the attention of the reader to the position occupied in the general system of geometry by certain propositions, which are in a certain sense *hypotheses equivalent to the Fifth Postulate.*

That our argument may be properly understood, we begin by explaining the meaning of this equivalence.

Two hypotheses are *absolutely equivalent* when each of them can be deduced from the other without the help of any new hypothesis. In this sense the two following hypotheses are absolutely equivalent:

a) Two straight lines parallel to a third are parallel to each other;

b) Through a point outside a straight line one and only one parallel to it can be drawn.

This kind of equivalence has not much interest, since the two hypotheses are simply two different forms of the same proposition. However we must consider in what way the idea of equivalence can be generalised.

Let us suppose that a deductive science is founded upon a certain set of hypotheses, which we will denote by $\{A, B, C, \ldots H\}$. Let M and N be two new hypotheses such that N can be deduced from the set $\{A, B, C \ldots H, M\}$, and M from the set $\{A, B, C \ldots H, N\}$

We indicate this by writing

$$\{A, B, C \ldots H, M\} \,.)\,.\, N,$$

and
$$\{A, B, C \ldots H, N\} .). M.$$

We shall now extend the idea of equivalence and say that the two hypotheses M, N are equivalent *relatively to the fundamental set* $\{A, B, C \ldots H\}$.

It has to be noted that the fundamental set $\{A, B, C \ldots H\}$ has an important place in this definition. Indeed it might happen that by diminishing this fundamental set, leaving aside, for example, the hypothesis A, the two deductions
$$\{B, C, \ldots H, M\} .). N$$
and
$$\{B, C, \ldots H, N\} .). M$$
could not hold *simultaneously*.

In this case the hypotheses M, N are *not equivalent* with respect to the new fundamental set $\{B, C \ldots H\}$.

After these explanations of a logical kind, let us see what follows from the discussion in the preceding chapters as to the equivalence between such hypotheses and the *Euclidean hypothesis*.

We assume in the first place as fundamental set of hypotheses that formed by the postulates of Association (A), and of Distribution (B), which characterise in the ordinary way the conceptions of the straight line and the plane: also by the postulates of Congruence (C), and the Postulate of Archimedes (D).

Relative to this fundamental set, which we shall denote by $\{A, B, C, D\}$, the following hypotheses are mutually equivalent, and equivalent also to that stated by EUCLID in his Fifth Postulate:

a) The internal angles, which two parallels make with a transversal on the same side, are supplementary [Ptolemy].

b) Two parallel straight lines are equidistant.

c) If a straight line intersects one of two parallels, it also intersects the other (Proclus);

or,

Two straight lines, which are parallel to a third, are parallel to each other;
or again,

Through a point outside a straight line there can be drawn one and only one parallel to that line.

d) A triangle being given, another triangle can be constructed similar to the given one and of any size whatever. [WALLIS.]

e) Through three points, not lying on a straight line, a sphere can always be drawn. [W. BOLYAI.]

f) Through a point between the lines bounding an angle a straight line can always be drawn which will intersect these two lines. [LORENZ.]

α) If the straight line r is perpendicular to the transversal AB and the straight line s cuts it at an acute angle, the perpendiculars from the points of s upon r are less than AB, on the side in which AB makes an acute angle with s. [NASÎR-EDDÎN.]

β) The locus of the points which are equidistant from a straight line is a straight line.

γ) The sum of the angles of a triangle is equal to two right angles. [SACCHERI.]

Now let us suppose that we diminish the fundamental set of hypotheses, *cutting out the Archimedean Hypothesis.* Then the propositions (a), (b), (c), (d), (e) and (f) are mutually equivalent, and also equivalent to the *Fifth Postulate of Euclid*, with respect to the fundamental set $\{A, B, C\}$. With regard to the propositions (α), (β), (γ), while they are mutually equivalent with respect to the set $\{A, B, C\}$ *no one* of them is equivalent to the *Euclidean Postulate.* This result brings out clearly the importance of the *Postulate of Archimedes.* It is given in the memoir of DEHN [1] [1900] to which

----

1 Cf. Note on p. 30.

reference has already been made. In that memoir it is shown
that the hypothesis (γ) on the sum of the angles of a triangle
is compatible not only with the ordinary elementary geometry, but also with a new geometry—necessarily Non-Archimedean—where the Fifth Postulate does not hold, and in
which an infinite number of lines pass through a point and
do not intersect a given straight line. To this geometry the
author gave the name of *Semi-Euclidean Geometry*.

The Spread of Non-Euclidean Geometry.

§ 60. The works of LOBATSCHEWSKY and BOLYAI did
not receive on their publication the welcome which so many
centuries of slow and continual preparation seemed to
promise. However this ought not to surprise us. The
history of scientific discovery teaches that every radical change
in its separate departments does not suddenly alter the con-
victions and the presuppositions upon which investigators
and teachers have for a considerable time based the present-
ation of their subjects.

In our case the acceptance of the Non-Euclidean Geometry was delayed by special reasons, such as the difficulty
of mastering LOBATSCHEWSKY's works, written as they were in
Russian, the fact that the names of the two discoverers were
new to the scientific world, and the Kantian conception of
space which was then in the ascendant.

LOBATSCHEWSKY's French and German writings helped
to drive away the darkness in which the new theories were
hidden in the first years; more than all availed the constant
and indefatigable labors of certain geometers, whose names
are now associated with the spread and triumph of Non-
Euclidean Geometry. We would mention particularly: C. L.
GERLING [1788—1864], R. BALTZER [1818—1887] and FR.
SCHMIDT [1827—1901], in Germany; J. HOÜEL [1823—

1886], G. BATTAGLINI [1826—1894], E. BELTRAMI [1835—
1900], and A. FORTI, in France and Italy.

§ 61. From 1816 GERLING kept up a correspondence
upon parallels with GAUSS[1], and in 1819 he sent him
SCHWEIKART's memorandum on *Astralgeometrie* [cf. p. 75].
Also he had heard from GAUSS himself [1832], and in terms
which could not help exciting his natural curiosity, of a
kleine Schrift on Non-Euclidean Geometry written by a
young Austrian officer, son of W. BOLYAI.[2] The bibliographical notes he received later from GAUSS [1844] on the works
of LOBATSCHEWSKY and BOLYAI[3] induced GERLING to procure
for himself the *Geometrischen Untersuchungen* and the *Appendix,* and thus to rescue them from the oblivion into which
they seemed plunged.

§ 62. The correspondence between GAUSS and SCHU-
MACHER, published between 1860 and 1863,[4] the numerous
references to the works of LOBATSCHEWSKY and BOLYAI, and
the attempts of LEGENDRE to introduce even into the elementary text books a rigorous treatment of the theory of parallels, led BALTZER, in the second edition of his *Elemente der*

---

[1] Cf. GAUSS, *Werke,* Bd. VIII, p. 167—169.

[2] Cf. GAUSS's letter to GERLING (GAUSS, *Werke,* Bd. VIII,
p. 220). In this note GAUSS says with reference to the contents
of the *Appendix:* "*worin ich alle meine eigenen Ideen und Resultate
wiederfinde mit großer Eleganz entwickelt.*" And of the author of
the work: „*Ich halte diesen jungen Geometer v. Bolyai für ein Genie
erster Größe*".

[3] Cf. GAUSS, *Werke,* Bd. VIII, p. 234—238.

[4] *Briefwechsel zwischen C. F. Gauss und H. C. Schumacher,*
Bd. II, p. '268—431 Bd. V, p. 246 (Altona, 1860—1863). As to
GAUSS's opinions at this time, see also, SARTORIUS VON WALTERS-
HAUSEN, *Gauß zum Gedächtnis,* p. 80—81 (Leipzig, 1856). Cf. GAUSS,
Werke, Bd. VIII, p. 267—268.

Mathematik (1867), to substitute, for the Euclidean definition of parallels one derived from the new conception of space. Following LOBATSCHEWSKY he placed the equation $A + B + C = 180°$, which characterises the Euclidean triangle, among the experimental results. To justify this innovation, BALTZER did not fail to insert a brief reference to the possibility of a more general geometry than the ordinary one, founded on the hypothesis of two parallels. He also gave suitable prominence to the names of its founders.[1] At the same time he called the attention of HOÜEL, whose interest in the question of elementary geometry was well known to scientific men,[2] to the Non-Euclidean geometry, and requested him to translate the *Geometrischen Untersuchungen* and the *Appendix* into French.

§ 63. The French translation of this little book by LOBATSCHEWSKY appeared in 1866 and was accompanied by some extracts from the correspondence between GAUSS and SCHUMACHER.[3] That the views of LOBATSCHEWSKY, BOLYAI, and GAUSS were thus brought together was extremely fortunate, since the name of GAUSS and his approval of the discoveries of the two geometers, then obscure and unknown,

---

[1] Cf. BALTZER, *Elemente der Mathematik*, Bd. II (5. Auflage) p. 12—14 (Leipzig, 1878). Also T. 4, p. 5—7, of CREMONA's translation of that work (Genoa, 1867).

[2] In 1863 HOÜEL had published his wellknown *Essai d'une exposition rationelle des principes fondamentaux de la Géométrie élémentaire*. Archiv d. Math. u. Physik, Bd. XL (1863).

[3] Mém. de la Soc. des Sci. de Bordeaux, T. IV, p. 88—120 (1866). This short work was also published separately under the title *Études géométriques sur la théorie des parallèles* par N. I. LOBATSCHEWSKY, Conseiller d'État de l'Empire de Russie et Professeur à l'Université de Kasan: traduit de l'allemand par J. HOÜEL, *suivie d'un Extrait de la correspondance de Gauss et de Schumacher*, (Paris, G. VILLARS, 1866).

helped to bring credit and consideration to the new doctrines in the most efficacious and certain manner.

The French translation of the *Appendix* appeared in 1867.[1] It was preceded by a *Notice sur la vie et les travaux des deux mathématiciens hongrois W. et J. Bolyai de Bolya,* written by the architect Fr. SCHMIDT at the invitation of HOÜEL,[2] and was supplemented by some remarks by W. BOL-YAI, taken from Vol. I of the *Tentamen* and from a short analysis, also by WOLFGANG, of the Principles of Arithmetic and Geometry.[3]

In the same year [1867] SCHMIDT's discoveries regarding the BOLYAIS were published in the *Archiv d. Math. u. Phys.* Also in the following year A. FORTI, who had already written a critical and historical memoir on LOBATSCHEWSKY,[4]

---

[1] Mém. de la Soc. des Sc. de Bordeaux, T. V, p. 189—248. This short work was also published separately unter the title: *La Science absolute de l'espace, indépendante de la vérité ou fausseté de l'Axiome XI d'Euclide (que l'on ne pourra jamais établir a priori); suivie de la quadrature géometrique du cercle, dans le cas de la fausseté de l'Axiome XI,* par Jean Bolyai, Capitaine au Corps du génie dans l'armée autrichienne; *Précédé d'une notice sur la vie et les travaux de W. et de J. Bolyai,* par M. FR. SCHMIDT, (Paris, G. VILLARS, 1868).

[2] Cf. P. STÄCKEL, *Franz Schmidt,* Jahresber. d. Deutschen Math. Ver., Bd. XI, p. 141—146 (1902).

[3] This little book of W. BOLYAI's is usually referred to shortly by the first words of the title *Kurzer Grundriss.* It was published at Maros-Vásárhely in 1851.

[4] *Intorno alla geometria immaginaria o non euclidiana. Considerazioni storico-critiche.* Rivista Bolognese di scienze, lettere, arti e scuole, T. II, p. 171—184 (1867). It was published separately as a pamphlet of 16 pages (Bologna, Fava e Garagnani, 1867). The same article, with some additions and the title, *Studii geometrici sulla teorica delle parallele di* N. J. Lobatschewky, appeared in the political journal *La Provincia di Pisa,* Anno III, Nr. 25, 27,

made the name and the works of the two now celebrated Hungarian geometers known to the Italians.[1]

To the credit of HoÜEL there should also be mentioned his interest in the manuscripts of JOHANN BOLYAI, then [1867] preserved, in terms of WOLFGANG's will, in the library of the Reformed College of Maros-Vásárhely. By the help of Prince B. BONCAMPAGNI [1821—1894], who in his turn interested the Hungarian Minister of Education, Baron EÖTVÖS, he succeeded in having them placed [1869] in the Hungarian Academy of Science at Budapest.[2] In this way they became more accessible and were the subject of painstaking and careful research, first by SCHMIDT and recently by STÄCKEL.

In addition HoÜEL did not fail in his efforts, on every available opportunity, to secure a lasting triumph for the Non-Euclidean Geometry. If we simply mention his *Essai critique sur les principes fondamenteaux de la géométrie:*[3] his article, *Sur l'impossibilité de démontrer par une construction plane le postulatum d'Euclide;*[4] the *Notices sur la vie et les travaux de N. J. Lobatschewsky;*[5] and finally his translations of various writings upon Non-Euclidean Geometry into French,[6]

29, 30 (1867); and part of it was reprinted under the original title (Pisa, Nistri, 1867).

[1] Cf. *Intorno alla vita ed agli scritti di Wolfgang e Giovanni Bolyai di Bolya, matematici ungheresi.* Boll. di Bibliografia e di Storia delle Scienze Mat. e Fisiche. T. I, p. 277—299 (1869). Many historical and bibliographical notes were added to this article of Forti's by B. BONCOMPAGNI.

[2] Cf. STÄCKEL's article on *Franz Schmidt* referred to above.

[3] 1. Ed., G. VILLARS, Paris, 1867; 2 Ed., 1883 (cf. Note 3 p. 52).

[4] Giornale di Mat. T. VII p. 84—89; Nouvelles Annales (2) T. IX, p. 93—96.

[5] Bull. des. Sc. Math. T. I, p. 66—71, 324—328, 384—388 (1870).

[6] In addition to the translations mentioned in the text, HoÜEL

it will e understood how fervent an apostle this science had found in the famous French mathematician.

HoÜEL's labours must have urged J. FRISCHAUF to perform the service for Germany which the former had rendered to France. His book—*Absolute Geometrie nach J. Bolyai* — (1872)[1] is simply a free translation of JOHANN's *Appendix*, to which were added the opinions of W. BOLYAI on the Foundations of Geometry. A new and revised edition of FRISCH-AUF's work was brought out in 1876[2]. In that work reference is made to the writings of LOBATSCHEWSKY and the memoirs of other authors who about that time had taken up this study from a more advanced point of view. This volume remained for many years the only book in which these new doctrines upon space were brought together and compared.

§ 64. With equal conviction and earnestness GIUSEPPE BATTAGLINI introduced the new geometrical speculations into Italy and there spread them abroad. From 1867 the *Giornale di Matematica*, of which he was both founder and editor, became the recognized organ of Non-Euclidean Geometry.

BATTAGLINI's first memoir—*Sulla geometria immaginaria di Lobatschewsky*[3]—was written to establish directly the principle which forms the foundation of the general theory of parallels and the trigonometry of LOBATSCHEWSKY. It was

---

translated a paper by BATTAGLINI (cf. note 3), two by BELTRAMI (cf. note 2 p. 127 and p. 147); one, by RIEMANN (cf. note p. 138), and one by HELMHOLTZ (cf. note p. 152).

[1] (xii + 96 pages) (Teubner, Leipzig).

[2] *Elemente der Absoluten Geometrie*, (vi + 142 pages) (Teubner, Leipzig).

[3] Giornale di Mat. T. V, p. 217—231 (1867). Rend. Acc. Science Fis. e Matem. Napoli, T. VI, p. 157—173 (1867). French translation, by HoÜEL, Nouvelles Annales (2) T. VII, p. 209—21, 265—277 (1868).

followed, a few pages later, by the Italian translation of the *Pangéométrie*[1]; and this, in its turn, in 1868, by the translation of the *Appendix*.

At the same time, in the sixth volume of the *Giornale di Matematica*, appeared E. BELTRAMI's famous paper, *Saggio di interpretazione della geometria non euclidea*.[2] This threw an unexpected light on the question then being debated regarding the fundamental principles of geometry, and the conceptions of GAUSS and LOBATSCHEWSKY.[3]

Glancing through the subsequent volumes of the *Giornale di Matematica* we frequently come upon papers upon Non-Euclidean Geometry. There are two by BELTRAMI [1872] connected with the above—named *Saggio;* several by BATTAGLINI [1874—78] and by d'OVIDIO [1875—77], which treat some questions in the new geometry by the projective methods discovered by CAYLEY; HOÜEL's paper [1870] on the impossibility of demonstrating Euclid's Postulate; and others by CASSANI [1873—81], GÜNTHER [1876], DE ZOLT [1877], FRATTINI [1878], RICORDI [1880], etc.

§ 65. The work of spreading abroad the knowledge of the new geometry, begun and energetically carried forward by the aforesaid geometers, received a powerful impulse from another set of publications, which appeared about this time [1868—72]. These regarded the problem of the foundations of geometry in a more general and less elementary way than that which had been adopted in the investigations of GAUSS,

---

[1] This was also published separately as a small book, entitled, *Pangeometria o sunto di geometria fondata sopra una teoria generale e rigorosa delle parallele* (Naples, 1867; 2a Ed. 1874).

[2] It was translated into French by HOÜEL in the Ann. Sc. de l'École Normale Sup., T. VI, p. 251—288 (1869).

[3] Cf. *Commemorazione di E. Beltrami* by L. CREMONA: Giornale di Mat. T. XXXVIII, p. 362 (1900). Also the *Nachruf* by E. PASCAL, Math. Ann. Bd. LVII, p. 65—107 (1903).

LOBATSCHEWSKY, and BOLYAI. In Chapter V. we shall shortly
describe these new methods and developments, which are associated with the names of some of the most eminent mathematicians and philosophers of the present time. Here it is
sufficient to remark that the old question of parallels, from
which all interest seemed to have been taken by the investigations of LEGENDRE forty years earlier, once again and
under a completely new aspect attracted the attention of geometers and philosophers, and became the centre of an
extremely wide field of labour. Some of these efforts were
simply directed toward rendering the works of the founders
of Non-Euclidean geometry more accessible to the general
mathematical public. Others were prompted by the hope of
extending the results, the content, and the meaning of the
new doctrines, and at the same time contributing to the progress of certain special branches of Higher Mathematics.[1]

---

[1] Cf. e. g., É. PICARD, La Science Moderne et son état
actuel, p. 75 (Paris, FLAMMARION, 1905).

Chapter V.

The Later Development of Non-Euclidean Geometry.

§ 66. To describe the further progress of Non-Euclidean Geometry in the direction of *Differential Geometry* and *Projective Geometry*, we must leave the field of Elementary Mathematics and speak of some of the branches of Higher Mathematics, such as the *Differential Geometry of Manifolds*, the *Theory of Continuous Transformation Groups*, *Pure Projective Geometry* (the system of STAUDT) and the *Metrical Geometries* which are subordinate to it. As it is not consistent with the plan of this work to refer, even shortly, to these more advanced questions, we shall confine ourselves to those matters without which the reader could not understand the motive spirit of the new researches, nor be led to that other geometrical system, due to RIEMANN, which has been altogether excluded from the previous investigations, as they assume that the straight line is of infinite length.

This system is known by the name of its discoverer and corresponds to the *Hypothesis of the Obtuse Angle* of SACCHERI and LAMBERT.[1]

---

[1] The reader, who wishes a complete discussion of the subject of this chapter, should consult KLEIN's *Vorlesungen über die nichteuklidische Geometrie*, (Göttingen, 1903); and BIANCHI's *Lezioni sulla Geometria differenziale*, 2 Ed. T. I, Cap. XI−XIV (Pisa, Spoerri, 1903). German translation by LUKAT, 1st Ed. (Leipzig, 1899). Also *The Elements of Non-Euclidean Geometry* by T. L. COOLIDGE which has recently (1909) been published by the Oxford University Press.

Differential Geometry and Non-Euclidean Geometry.

The Geometry upon a Surface.

§ 67. What follows will be more easily understood if we start with a few observations:

A surface being given, let us see how far we can establish a geometry upon it analogous to that on the plane.

Through two points A and B on the surface there will generally pass one definite line belonging to the surface, namely, the shortest distance on the surface between the two points. This line is called the *geodesic* joining the two points. In the case of the sphere, the geodesic joining two points, not the extremities of a diameter, is an arc of the great circle through the two points.

Now if we wish to compare the geometry upon a surface with the geometry on a plane, it seems natural to make the geodesics, which measure the distances on the one surface, correspond to the straight lines of the other. It is also natural to consider two figures traced upon the surface as (*geodetically*) *equal*, when there is a point to point correspondence between them, such that the geodesic distances between corresponding points are equal.

We obtain a representation of this conception of equality if we assume that the surface is made of a *flexible* and *inextensible* sheet. Then by a movement of the surface, which does not remain rigid, but is bent as described above, those figures upon it, which we have called equal, are to be superposed the one upon the other.

Let us take, for example, a piece of a cylindrical surface. By simple bending, without stretching, folding, or tearing, this can be *applied* to a plane area. It is clear that in this case two figures ought to be called equal *on the surface*, which coincide with equal areas on the plane, though of course two such figures are not in general equal in space.

Returning now to any surface whatsoever, the system of conventions, suggested above, leads to a *geometry on the surface*, which we propose to consider always for suitably bounded regions [*Normal Regions*]. Two surfaces which are applicable the one to the other, by bending without stretching, will have the same geometry. Thus, for example, upon any cylindrical surface whatsoever, we will have a geometry similar to that on any plane surface, and, in general, upon any *developable* surface.

The geometry on the sphere affords an example of a geometry on a surface essentially different from that on the plane, since it is impossible to apply a portion of the sphere to the plane. However there is an important analogy between the geometry on the plane and the geometry on the sphere. This analogy has its foundation in the fact that the sphere can be freely moved upon itself, so that propositions in every way analogous to the postulates of congruence on the plane hold for equal figures on the sphere.

Let us try to generalize this example. In order that a suitably bounded surface, by bending but without stretching, can be moved upon itself in the same way as a plane, a certain number [*K*], invariant with respect to this bending, must have a *constant value* at all points of the surface. This number was introduced by Gauss and called the *Curvature*.[1] [In English books it is usually called *Gauss's Curvature* or the *Measure of Curvature*.]

---

[1] Remembering that the curvature at any point of a plane curve is the reciprocal of the radius of the osculating circle for that point, we shall now show that the curvature at a point *M* of the surface can be defined. Having drawn the normal *n* to the surface at *M*, we consider the pencil of planes through *n*, and the corresponding pencil of curves formed by their intersections with the surface. In this pencil of (plane) curves, there are two, orthogonal to each other, whose curvatures, as defined above, are maximum and minimum. The product of their curvatures is Gauss's Curvature of the Surface at M. This Curvature has one most marked

Surfaces of Constant Curvature can be actually constructed. The three cases

$$K = 0, \quad K > 0, \quad K < 0,$$

have to be distinguished.

For $K = 0$, we have the developable surfaces [applicable to the plane].

For $K > 0$, we have the surfaces applicable to a sphere of radius $1 : \sqrt{K}$, and the sphere can be taken as a *model* for these surfaces.

For $K < 0$, we have the surfaces applicable to the *Pseudosphere*, which can be taken as a model for the surfaces of *constant negative curvature*.

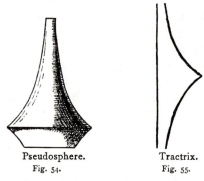

Pseudosphere.   Tractrix.
Fig. 54.   Fig. 55.

The Pseudosphere is a surface of revolution. The equation of its meridian curve (the tractrix [1]) referred to the axis

---

characteristic. It is unchanged for every bending of the surface which does not involve stretching. Thus, if two surfaces are applicable to each other in the sense of the text, they ought to have the same Gaussian Curvature at corresponding points [GAUSS].

This result, the converse of which was proved by MINDING to hold for Surfaces of Constant Curvature, shows that surfaces, freely movable upon themselves, are characterised by constancy of curvature.

[1] The tractrix is the curve in which the distance from the

of rotation z, and to a suitably chosen axis of x perpendicular to z, is

(1) $$z = k \log \frac{k + \sqrt{k^2 - x^2}}{x} - \sqrt{k^2 - x^2},$$

where k is connected with the Curvature K by the equation

$$K = -\frac{1}{k^2}.$$

To the pseudosphere generated by (1) can be applied any portion of the surface of constant curvature $-\frac{1}{k^2}$.

Surface of Constant Negative Curvature.[1]

Fig. 56.

point of contact of a tangent to the point where it cuts its asymptote is constant.

[1] Fig. 56 is reproduced from a photograph ef a surface constructed by BELTRAMI. The actual model belongs to the collection of models in the Mathematical Institute of the University of Pavia.

§ 68.   There is an analogy between the geometry on a surface of constant curvature and that of a portion of a plane, both taken within suitable boundaries. We can make this analogy clear by *translating* the fundamental definitions and properties of the one into those of the other. This is indicated shortly by the positions which the corresponding terms occupy in the following table:

(a) Surface.	(a) Portion of the plane.
(b) Point.	(b) Point.
(c) Geodesic.	(c) Straight line.
(d) Arc of Geodesic.	(d) Rectilinear Segment.
(e) Linear properties of the Geodesic.	(e) Postulates of Order for points on a Straight Line.
(f) A Geodesic is determined by two points.	(f) A Straight Line is determined by two points.
(g) Fundamental properties of the equality of Geodesic Arcs and Angles.	(g) Postulates of Congruence for Rectilinear Segments and Angles.
(h) If two Geodesic triangles have their two sides and the contained angles equal, then the remaining sides and angles are equal.	(h) If two Rectilinear triangles have their two sides and the contained angles equal, then the remaining sides and angles are equal.

It follows that we can retain as common to the geometry of the said surfaces all those properties concerning bounded regions on a plane, which in the Euclidean system are independent of the Parallel Postulate, when no use is made of the *complete plane* [e. g., of the infinity of the straight line] in their demonstration.

We must now proceed to compare the propositions for a bounded region of the plane, depending on the Euclidean hypothesis, with those which correspond to them in the geometry on the surface of constant curvature. We have, e. g., the proposition that the sum of the angles of a triangle is equal to two right angles. The corresponding property does not generally hold for the surface.

Indeed GAUSS showed that upon a surface whose curvature K is constant or varies from point to point, the surface integral

$$\iint K dS,$$

over the whole surface of a geodesic triangle ABC, is *equal to the excess of its three angles over two right angles.*[1]

i. e.
$$\iint_{ABC} K dS = A + B + C - \pi.$$

Let us apply this formula to the surfaces of constant curvature, distinguishing the three possible cases—

Case I. $\qquad\qquad K = 0.$

In this case we have

$$\iint_{ABC} K dS = 0; \text{ that is } A + B + C = \pi.$$

Thus *the sum of the angles of a geodesic triangle on surfaces of zero curvature is equal to two right angles.*

Case II. $\qquad\qquad K = \frac{1}{k^2} > 0.$

In this case we have

$$\iint_{ABC} K dS = \frac{1}{k^2} \iint_{ABC} dS.$$

But $\iint dS =$ area of the triangle $ABC = \Delta.$

$$\therefore \quad \frac{\Delta}{k^2} = A + B + C - \pi.$$

From this equation it follows that
$$A + B + C > \pi,$$
and that $\qquad \Delta = k^2 (A + B + C - \pi).$

---

[1] Cf. BIANCHI's work referred to above; Chapter VI.

That is:

a) *The sum of the angles of a geodesic triangle on sur-faces of constant positive curvature is greater than two right angles.*

b) *The area of a geodesic triangle is proportional to the excess of the sum of its angles over two right angles.*

Case III. $K = -\dfrac{1}{k^2} < 0$

In this case we have

$$\iint\limits_{ABC} K dS = -\frac{1}{k^2} \iint\limits_{ABC} dS = -\frac{\Delta}{k^2},$$

where we again denote the area of the triangle ABC by Δ.

Then we have

$$\frac{\Delta}{k^2} = \pi - (A + B + C).$$

From this it follows that

$$A + B + C < \pi,$$

and that $\Delta = k^2 (\pi - A - B - C).$

That is:

a) *The sum of the angles of a geodesic triangle on sur-faces of constant negative curvature is less than two right angles.*

b) *The area of a geodesic triangle is proportional to the difference between the sum of its angles and two right angles.*

We bring these results together in the following table:

Surfaces of Constant Curvature.

Value of the Curvature	Model of the Surface	Character of the Triangle
$K = 0$	Plane	$\measuredangle A + \measuredangle B + \measuredangle C = \pi$
$K = \dfrac{1}{k^2}$	Sphere	$\measuredangle A + \measuredangle B + \measuredangle C > \pi$
$K = -\dfrac{1}{k^2}$	Pseudosphere	$\measuredangle A + \measuredangle B + \measuredangle C < \pi$

With the geometry of surfaces of zero curvature and of surfaces of constant positive curvature we are already acquainted, since they correspond to Euclidean plane geometry and to spherical geometry.

The study of the surfaces of constant negative curvature was begun by F. MINDING [1806—1885] with the investigation of the surfaces of revolution to which they could be applied.[1] The following remark of MINDING's, fully proved by D. CODAZZI [1824—1873], establishes the trigonometry óf such surfaces. *In the formulae of spherical trigonometry let the angles be kept fixed and the sides multiplied by* $i = \sqrt{-1}$. *Then we obtain the equations which are satisfied by the elements of the geodesic triangles on the surfaces of constant negative curvature.*[2] These equations [the *pseudospherical trigonometry*] evidently coincide with those found by TAURINUS; in other words, with the formulæ of the geometry of LOBATSCHEWSKY-BOLYAI.

§ 69. From the preceding paragraphs it will be seen that the theorems regarding the sum of the angles of a triangle in the geometry on surfaces of constant curvature, are related to those of plane geometry as follows:—

For $K = O$ they correspond to those which hold on the plane in the case of the *Hypothesis of the Right Angle.*

For $K > O$ they correspond to those which hold on the plane in the case of the *Hypothesis of the Obtuse Angle.*

---

[1] *Wie sich entscheiden lässt, ob zwei gegebene krumme Flächen aufeinander abwickelbar sind oder nicht; nebst Bemerkungen über die Flächen von unveränderlichem Krümmungsmasse.* CRELLE's Journal, Bd. XIX, p. 370—387 (1839).

[2] MINDING: *Beiträge zur Theorie der kürzesten Linien auf krummen Flächen.* CRELLE's Journal, Bd. XX, p. 323—327 (1840). D. CODAZZI: *Intorno alle superficie, le quali hanno costante il prodotto de' due raggi di curvatura.* Ann. di Scienze Mat. e Fis. T. VIII, p. 346—355 (1857).

For $K < O$ they correspond to those which hold on the plane in the case of the *Hypothesis of the Acute Angle.*

The first of the results is evident a priori, since we are concerned with developable surfaces.

The analogy between the geometry of the surfaces of constant negative curvature, for example, and the geometry of LOBATSCHEWSKY-BOLYAI, could be made still more evident by arranging in tabular form the relations between the elements of the geodesic triangles traced upon those surfaces, and the formulæ of Non-Euclidean Trigonometry. Such a comparison was made by E. BELTRAMI in his *Saggio di interpretazione della geometria non-euclidea.*[1]

In this way it will be seen that the geometry upon a surface of constant positive or negative curvature can be considered as a *concrete interpretation of the Non-Euclidean Geometry, obtained in a bounded plane area, with the aid of the Hypothesis of the Obtuse Angle or that of the Acute Angle.*

The possibility of interpreting the geometry of a two-dimensional manifold by means of ordinary surfaces was observed by B. RIEMANN [1826—1866] in 1854, the year in which he wrote his celebrated memoir: *Über die Hypothesen welche der Geometrie zugrunde liegen.*[2] The developments of

---

[1] Giorn. di Mat., T. VI, p. 284—312 (1868).   Opere Mat., T. I, p. 374—405 (Hoepli, Milan, 1902).

[2] *Riemanns Werke*, 1. Aufl. (1876), p. 254—312: 2. Aufl. (1892), p. 272—287. It was read by RIEMANN to the Philosophical Faculty at Göttingen as his *Habilitationsschrift*, before an audience not composed solely of mathematicans. For this reason it does not contain analytical developments, and the conceptions introduced are mostly of an intuitive character. Some analytical explanations are to be found in the notes on the Memoir sent by RIEMANN as a solution of a problem proposed by the Paris Academy (*Riemanns Werke*, 1. Aufl., p. 384—391). The philosophical basis of the *Habilitationsschrift* is the study of the properties of things from their behaviour as infinitesimals. Cf. KLEIN's discourse:

Non-Euclidean Geometry in the direction of Differential Geometry are directly due to this memoir.

BELTRAMI's interpretation appears as a particular case of RIEMANN's. It shows clearly, from the properties of surfaces of constant curvature, that the chain of deductions from the three hypotheses regarding the sum of the angles of a triangle must lead to logically consistent systems of geometry.

This conclusion, so far as regards the *Hypothesis of the Obtuse Angle*, seems to contradict the theorems of SACCHERI, LAMBERT, and LEGENDRE, which altogether exclude the possibility of a geometry founded on that hypothesis. However the contradiction is only apparent. It disappears if we remember that in the demonstration of these theorems, not only the fundamental properties of the bounded plane are used, but also those of the complete plane, e. g., the property that the straight line is infinite.

Principles of Plane Geometry on the Ideas of Riemann.

§ 70. The preceding observations lead us to the foundation of a metrical geometry, which excludes Euclid's Postul-

---

Riemann und seine Bedeutung in der Entwickelung der modernen Mathematik. Jahresb. d. Deutschen Math. Ver., Bd. IV, p. 72—82 (1894), and the Italian translation by E. PASCAL in Ann. di Mat., (2), T. XXIII, p. 222. The *Habilitationsschrift* was first published in 1867 after the death of the author [Gött. Abh. XIII] under the editorship of DEDEKIND. It was then translated into French by J. HOÜEL [Ann. di Mat. (2). T. III (1870), Oeuvres de Riemann, (1876)]; into English, by W. K. CLIFFORD [Nature, Vol. VIII, (1873)], and again by G. B. HALSTED [Tokyo sagaku butsurigaku kwai kiji, Vol. VII, (1895); into Polish, by DICKSTEIN (Comm. Acad. Litt. Cracov. Vol. IX, 1877); into Russian, by D. SINTSOFF [Mem. of the Physical Mathematical Society of the University of Kasan, (2), Vol. III, App. (1893)].

ate, and adopts a more general point of view than that formerly held:

(*a*) *We assume that we start from a bounded plane area (normal region), and not from the whole plane.*

(*b*) *We regard as postulates those elementary propositions, which are revealed to us by the senses for the region originally taken; the propositions relative to the straight line being determined by two points, to congruence, etc.*

(*c*) *We assume that the properties of the initial region can be extended to the neighbourhood of any point on the plane [we do not say to the complete plane, viewed as a whole].*

The geometry, built upon these foundations, will be the most general plane geometry, consistent with the data which rigorously express the result of our experience. These results are, however, limited to an accessible region.

From the remarks in § 69, it is clear that the said geometry will find a concrete interpretation in that of the surfaces of constant curvature.

This correspondence, however, exists only from the point of view (*differential*) according to which only bounded regions are compared. If, on the other hand, we place ourselves at the (*integral*) point of view, according to which the geometry of the whole plane and the geometry on the surface are compared, the correspondence no longer exists. Indeed, from this standpoint, we cannot even say that the same geometry will hold on two surfaces with the same constant curvature. For example, a circular cylinder has a constant curvature, zero, and a portion of it can be applied to a region of a plane, but the entire cylinder cannot be applied in this way to the entire plane. The geometry of the complete cylinder thus differs from that of the complete Euclidean plane. Upon the cylinder there are closed geodesics (its circular sections), and, in general, two of its geodesics (helices) meet in an infinite number of points, instead of in just two.

Similar differences will in general appear between a metrical Non-Euclidean geometry, founded on the postulates enunciated above, and the geometry on a corresponding surface of constant curvature.

When we attempt to consider the geometry on a surface of constant curvature (e. g., on the sphere or pseudosphere) as a whole, we see, in general, that the fundamental property of a normal region that a geodesic is fully determined by two points ceases to hold. This fact, however, is not a necessary consequence of the hypotheses on which, in the sense above explained, a general metrical Non-Euclidean geometry of the plane is based. Indeed, when we examine whether a system of plane geometry is logically possible, which will satisfy the conditions (a), (b), and (c), and in which the postulates of congruence and that a straight line is fully determined by two points are valid on the complete plane, we obtain, in addition to the ordinary Euclidean system, the two following systems of geometry:

1. The *system of Lobatschewsky-Bolyai*, already explained, in which two parallels to a straight line pass through a point.

2. *A new system* (called *Riemann's system*) which corresponds to SACCHERI's *Hypothesis of the Obtuse Angle*, and in which no parallel lines exist.

In the latter system the straight line is a *closed* line of finite length. We thus avoid the contradiction to which we would be led if we assumed that the straight line were *open* (infinite). This hypothesis is required in proving Euclid's Theorem of the Exterior Angle [I. 16] and some of SACCHERI's results.

§ 71. RIEMANN was the first to recognize the existence of a system of geometry compatible with the *Hypothesis of the Obtuse Angle*, since he was the first to substitute for the hypothesis that the straight line is *infinite*, the more general one that it is *unbounded*. The difference, which presents itself here, between *infinite* and *unbounded* is most important. We quote in regard to this RIEMANN's own words:[1]

'In the extension of space construction to the infinitely great, we must distinguish between *unboundedness* and *infinite extent;* the former belongs to the extent relations; the latter to the measure relations. That space is an unbounded three-fold manifoldness is an assumption which is developed by every conception of the outer world; according to which every instant the region of real perception is completed and the possible positions of a sought object are constructed, and which by these applications is for ever confirming itself. The unboundedness of space possesses in this way a greater empirical certainty than any external experience, but its infinite extent by no means follows from this; on the other hand, if we assume independence of bodies from position, and therefore ascribe to space constant curvature, it must necessarily be finite, provided this curvature has ever so small a positive value.'

Finally, the postulate which gives the straight line an infinite length, implicitly contained in the work of preceding geometers, is to RIEMANN as fit a subject of discussion as that of parallels. What RIEMANN holds as beyond discussion is the *unboundedness* of space. This property is compatible with the hypothesis that the straight line is infinite (open), as well as with the hypothesis that it is finite (closed).

The logical possibility of RIEMANN's system can be deduced from its concrete interpretation in *the geometry of the sheaf of lines*. The properties of the sheaf of lines are trans-

---

[1] [This quotation is taken from CLIFFORD's translation in Nature, referred to above. (Teil III, § 2 of RIEMANN's Memoir.)].

lated readily into those of RIEMANN's plane, and vice versa, with the aid of the following *dictionary:*

Sheaf	Plane
Line	Point
Plane [Pencil]	Straight line
Angle between two Lines	Segment
Dihedral Angle	Angle
Trihedron	Triangle
· · · · · ·	· · · · ·
· · · · · ·	· · · · ·

We now give, as an example, the 'translation' of some of the best known propositions for the sheaf:

a) The sum of the three dihedral angles of a trihedron is greater than two right dihedral angles.

a) The sum of the three angles of a triangle is greater than two right angles.

b) All the planes which are perpendicular to another plane pass through a straight line.

b) All the straight lines perpendicular to another straight line pass through a point.

c) With every plane of the sheaf let us associate the straight line in which the planes perpendicular to the given plane all intersect. In this way we obtain a correspondence between planes and straight lines which enjoys the following property: The straight lines corresponding to the planes of a pencil [Ebenenbüschel, set of planes through one line, the axis of the pencil] lie on a plane,

c) With every straight line in the plane let us associate the point in which the lines perpendicular to the given line intersect. In this way we obtain a correspondence between lines and points, which enjoys the following property:

The points corresponding to the lines of a pencil lie on a straight line, which in its turn has for corresponding point the vertex of the pencil.

which in its turn has for corresponding line the axis of
the pencil. The correspondence thus defined is called
absolute [orthogonal] *polarity*
of the sheaf.

The correspondence thus
defined is called *absolute polarity* of the plane.

§ 72. A remarkable discovery with regard to the *Hypothesis of the Obtuse Angle* was made recently by DEHN.

If we refer to the arguments of SACCHERI [p. 37],
LAMBERT [p. 45], LEGENDRE [p. 56], we see at once that
these authors, in their proof of the falsehood of the *Hypothesis of the Obtuse Angle*, avail themselves, not only of the
hypothesis that the straight line is infinite, but also of the
Archimedean Hypothesis. Now we might ask ourselves if this
second hypothesis is required in the proof of this result. In
other words, we might ask ourselves if the two hypotheses,
one of which attributes to the straight line the character of
open lines, while the other attributes to the sum of the angles
of a triangle a value greater than two right angles, are compatible with each other, when the *Postulate of Archimedes* is
excluded. DEHN gave an answer to this question in his
memoir quoted above (p. 30), by the construction of a *Non-
Archimedean* geometry, in which the straight line is open,
and the sum of the angles of a triangle is greater than two
right angles. Thus the second of SACCHERI's three hypotheses
is compatible with the hypothesis of the open straight line
in the sense of a *Non-Archimedean* system. This new
geometry was called by DEHN *Non-Legendrean Geometry* [cf.
§ 59, p. 121].

§ 73. We have seen above that the geometry of a
surface of constant curvature (positive or negative) does not
represent, in general, the whole of the Non-Euclidean geo-

metry on the plane of LOBATSCHEWKY and of RIEMANN. The question remains whether such a correspondence could not be effected with the help of some particular surface of this nature.

The answer to this question is as follows:

1) *There does not exist any regular*[1] *analytic surface on which the geometry of Lobatschewsky-Bolyai is altogether valid* [HILBERT's Theorem].[2]

---

[1] In other words, free from singularities.

[2] *Über Flächen von konstanter Gaussscher Krümmung.* Trans. Amer. Math. Soc. Vol. II, p. 86—99 (1901); Grundlagen der Geometrie, 2. Aufl. p. 162—175. (Leipzig, Teubner, 1903).

This question, which HILBERT's Theorem answers, was first suggested to mathematicians by BELTRAMI's interpretation of the LOBATSCHEWKY-BOLYAI Geometry. In 1870 HELMHOLTZ—in his lecture, *Über Ursprung und Bedeutung der geometrischen Axiome*, (Vorträge und Reden, Bd. II. Brunswick, 1844)—had denied the possibility of constructing a pseudospherical surface, extending indefinitely in every direction. Also A. GENNOCCHI—in his *Lettre à M. Quetelet sur diverses questions mathématiques*, [Belgique Bull. (2). T. XXXVI, p. 181—198 (1873)], and more fully in his Memoir, *Sur une mémoire de D. Foncenex et sur les géométries non-euclidiennes*, [Torino Memorie (3), T. XXIX, p. 365—404 (1877)], showed the insufficiency of some intuitive demonstrations, intended to prove the concrete existence of a surface suitable for the representation of the entire Non-Euclidean plane. Also he insisted upon the probable existence of singular points—(as for example, those on the line of regression of Fig. 54)—in every concrete model of a surface of constant negative curvature.

So far as regards HILBERT's Theorem, we add that the analytic character of the surface, assumed by the author, has been shown to be unnecessary. Cf. the dissertation of G. LÜTKEMEYER: *Über den analytischen Charakter der Integrale von partiellen Differentialgleichungen*, (Göttingen, 1902). Also the Note by E. HOLMGREN: *Sur les surfaces à courbure constante négative*, [Comptes Rendus, I Sem., p. 840—843 (1902)].

[In a recent paper *Sur les surfaces à courbure constante négative*, (Bull. Soc. Math. de France. t. XXXVII p. 51—58, 1909) É. GOURSAT

2) *A surface on which the geometry of the plane of Riemann would be altogether valid must be a closed surface.*

The only regular analytic closed surface of constant positive curvature is the sphere [LIEBMANN's Theorem].[1]

But on the sphere, in normal regions of which RIEMANN's geometry is valid, two lines always meet in two (opposite) points.

We therefore conclude that:

In ordinary space there are no surfaces which satisfy in their complete extent all the properties of the Non-Euclidean planes.

§ 74. At this place it is right to observe that the sphere, among all the surfaces whose curvature is constant and different from zero, has a characteristic that brings it nearer to the plane than all the others. Indeed the sphere can be moved upon itself just as the plane, so that the properties of congruence are valid not only for normal regions, but, as in the plane, for the surface of the sphere taken as a whole.

This fact suggests to us a method of enunciating the postulates of geometry, which does not exclude, a priori, the possible existence of a plane with all the characteristics of the sphere, including that of opposite points. We would

---

has discussed a problem slightly less general than that enunciated by HILBERT, and has succeeded in proving—in a fairly simple manner—the impossibility of constructing an analytical surface of constant curvature, which has no singular points at a finite distance.]

[1] *Eine neue Eigenschaft der Kugel*, Gött. Nachr. p. 44—54 (1899). This property is also proved by HILBERT on p. 172—175 of his *Grundlagen der Geometrie*. We notice that the surfaces of constant positive curvature are necessarily analytic. Cf. LÜTKEMEYER's Dissertation referred to above (p. 163), and the memoir by HOLMGREN: *Über eine Klasse von partiellen Differentialgleichungen der zweiten Ordnung*, Math. Ann. Bd. LVII, p. 407—420 (1903).

need to assume that the following relations were true for the plane:

1) The postulates (*b*), (*c*) [cf. § 70] in every normal region.

2) The postulates of congruence in the whole of the plane.

Thus we would have the geometrical systems of EUCLID, of LOBATSCHEWSKY-BOLYAI, and of RIEMANN (*the elliptic type*), which we have met above, where two straight lines have only one common point: and a second RIEMANN's system (*the spherical type*), where two straight lines have always two common points.

§ 75. We cannot be quite certain what idea RIEMANN had formed of his complete plane, whether he had thought of it as the *elliptic plane*, or the *spherical plane*, or had recognized the possibility of both. This uncertainty is due to the fact that in his memoir he deals with Differential Geometry and devotes only a few words to the complete forms. Further, those who continued his labours in this direction, among them BELTRAMI, always considered RIEMANN's geometry in connection with the sphere. They were thus led to hold that on the complete RIEMANN's plane, as on the sphere (owing to the existence of the opposite ends of a diameter), the postulate that a straight line is determined by two points had exceptions,[1] and that the only form of the plane compatible with the *Hypothesis of the Obtuse Angle* would be the *spherical plane*.

---

[1] Cf. for example, the short reference to the geometry of space of constant positive curvature with which BELTRAMI concludes his memoir: *Teoria fondamentale degli spazii di curvatura costante*, Ann. di Mat. (2). T. II, p. 354—355 (1868); or the French translation of this memoir by J. HOÜEL, Ann. Sc. d. l'École Norm. Sup. T. VI, p. 347—377.

The fundamental characteristics of the *elliptic plane* were given by A. CAYLEY [1821—1895] in 1859, but the connection between these properties and Non-Euclidean geometry was first pointed out by KLEIN in 1871. To KLEIN is also due the clear distinction between the two geometries of RIEMANN, and the representation of the elliptic geometry by the geometry of the sheaf [cf. § 71].

To make the difference between the spherical and elliptic geometries clearer, let us fix our attention on two classes of surfaces presented to us in ordinary space: the surface with two faces (*two-sided*) and the surface with one face (*one-sided*).

Examples of two-sided surfaces are afforded by the ordinary plane, the surfaces of the second order (conicoidal, cylindrical, and spherical), and in general all the surfaces enclosing solids. On these it is possible to distinguish two faces.

An example of a one-sided surface is given by the Leaf of MÖBIUS [MÖBIUSsche Blatt], which can be easily constructed as follows: Cut a rectangular strip *ABCD*. Instead of joining the opposite sides *AB* and *CD* and thus obtaining a cylindrical surface, let these sides be joined after one of them, e. g., *CD*, has been rotated through two right angles about its middle point. Then what was the upper face of the rectangle, in the neighbourhood of *CD*, is now succeeded by the lower face of the original rectangle.

Thus *on Möbius' Leaf the distinction between the two faces becomes impossible.*

If we wish to distinguish the one-sided surface from the wo-sided by a characteristic, depending only on the intrinsic properties of the surface, we may proceed thus:—We fix a point on the surface, and a direction of rotation about it. Then we let the point describe a closed path upon the surface, which does not leave the surface; for a two-sided surface the point returns to its initial position and the final direction of rotation coincides with the initial one; for a one-sided surface, [as can be easily verified on the Leaf of MÖBIUS, when the path coincides with the diametral line] there exist closed paths for which the final direction of rotation is opposite to the initial direction.

Coming back to the two RIEMANN's planes, we can now easily state in what their essential difference consists: *the spherical plane has the character of the two-sided surface, and the elliptic plane that of the one-sided surface.*

The Leaf of Möbius.
Fig. 57.

The property of the elliptic plane here enunciated, as well as all its other properties, finds a concrete interpretation in the sheaf of lines. In fact, if one of the lines of the sheaf is turned about the vertex through half a revolution, the two rotations which have this line for axis are interchanged.

Another property of the elliptic plane, allied to the preceding, is this: *The elliptic plane*, unlike the Euclidean plane and the other Non-Euclidean planes, *is not divided by its lines into two parts*. We can state this property otherwise: If two points A and A' are given upon the plane, and an arbitrary straight line, we can pass from A to A' by a path which does not leave the plane and does not cut the line.[1] This fact is 'translated' by an obvious property of the sheaf, which it would be superfluous to mention.

§ 76. The interpretation of the spherical plane by the *sheaf of rays* (straight lines starting from the vertex) is analogous to that given above for the elliptic plane. The trans-

---

[1] A surface which completely possesses the properties of the elliptic plane was constructed by W. BOY. [Gött. Berichte, p. 20 —23 (1900); Math. Ann. Bd. LVII, p. 151—184 (1903)].

lation of the properties of this plane into the properties of the sheaf of rays is effected by the use of a .'dictionary' similar to that of § 71, in which the word *point* is found opposite the word *ray*.

The comparison of the sheaf of rays with the sheaf of lines affords a useful means of making clear the connections, and revealing the differences, which are to be found in the two geometries of RIEMANN.

We can consider two sheaves, with the same vertex, the one of lines, the other of rays. It is clear that to every line of the first correspond two rays of the second; that every figure of the first is formed by two symmetrical figures of the second; and that, with certain restrictions, the metrical properties of the two forms are the same. Thus if we agree to regard the two opposite rays of the sheaf of rays as forming one element only, the sheaf of rays and the sheaf of lines are identical.

The same considerations apply to the two RIEMANN's planes. To every point of the elliptic plane correspond two distinct and opposite points of the spherical plane; to two lines of the first, which pass through that point, correspond two lines of the second, which have two points in common; etc.

The elliptic plane, when compared with the spherical plane, ought to be regarded as a *double plane.*

With regard to the elliptic plane and the spherical plane, it is right to remark that the formulæ of absolute trigonometry, given in § 56, can be applied to them in every suitably bounded region. This follows from the fact, already noted in § 58, that the formulæ of absolute trigonometry hold on the sphere, and the geometry of the sphere, so far as regards normal regions, coincides with that of these two planes.

Principles of Riemann's Solid Geometry.

§ 77. Returning now to solid geometry, we start from the philosophical foundation that the postulates, although we grant them, by hypothesis, an actual meaning, express truths of experience, which can be verified only in a bounded region. We also assume, that on the foundation of these postulates points in space are represented by three coordinates.

On such an (analytical) representation, every line is given by three equations in a single variable:

$$x_1 = f_1(t), \; x_2 = f_2(t), \; x_3 = f_3(t),$$

and we must now proceed to determine a function s, of the parameter t, which shall express the *length* of an arc of the curve.

On the strength of the *distributive property*, by which the length of an arc is equal to the sum of the lengths of the parts into which we imagine it to be divided, such a function will be fully determined when we know the *element of distance* (ds) between two infinitely near points, whose coordinates are

$$x_1, \; x_2, \; x_3,$$
$$x_1 + dx_1, \; x_2 + dx_2, \; x_3 + dx_3.$$

RIEMANN starts with very general hypotheses, which are satisfied most simply by assuming that ds^2, the square of the element of distance, is a quadratic expression involving the differentials of the variables, which always remains positive:

$$ds^2 = \Sigma a_{ij} \, dx_i \, dx_j,$$

where the coefficients a_{ij} are functions of $x_1, \, x_2, \, x_3$.

Then, admitting the principle of superposition of figures, it can be shown that the function a_{ij} must be such that, with the choice of a suitable system of coordinates,

$$ds^2 = \frac{dx_1^2 + dx_2^2 + dx_3^2}{1 + \dfrac{K}{4} \, (x_1^2 + x_2^2 + x_3^2)}$$

In this formula the constant K is what RIEMANN, by an extension of GAUSS's conception, calls the *Curvature of Space*.

According as K is greater than, equal to, or less than zero, we have space of constant positive curvature, space of zero curvature, or space of constant negative curvature.

We make another forward step when we assume that the principle of superposition [the principle of movement] can be extended to the whole of space, as also the postulate that a straight line is always determined by two points. In this way we obtain three forms of space; that is, three geometries which are logically possible, consistent with the data from which we set out.

The first of these geometries, corresponding to positive curvature, is characterised by the fact that RIEMANN's system is valid in every plane. For this reason space of positive curvature will be unbounded and finite in all directions. The second, corresponding to zero curvature, is the ordinary Euclidean geometry. And the third, which corresponds to negative curvature, gives rise in every plane to the geometry of LOBATSCHEWSKY-BOLYAI.

The Work of Helmholtz and the Investigations of Lie.

§ 78. In some of his philosophical and mathematical writings,[1] HELMHOLTZ [1821—1894] has also dealt with the

---

[1] *Über die thatsächlichen Grundlagen der Geometrie*, Heidelberg, Verh. d. naturw.-med. Vereins, Bd. IV, p. 197—202 (1868); Bd. V, p. 31—32 (1869). Wiss. Abhandlungen von H. HELMHOLTZ, Bd. II, p. 610—617 (Leipzig, 1883). French translation by J. HOÜEL in Mém. de la Soc. des Sc. Phys. et Nat. de Bordeaux, T. V, (1868), and also, in book form, along with the *Études Géométriques* of LOBATSCHEWSKY and the *Correspondance de Gauss et de Schumacher*, (Paris, Hermann, 1895).

Über die Thatsachen, die der Geometrie zum Grunde liegen. Gött.

question of the foundations of geometry. Instead of assuming a priori the form

$$ds^2 = \Sigma a_{ij}\, dx_i\, dx_j,$$

as the expression for the element of distance, he showed that this expression, in the form given to it by RIEMANN for space of constant curvature, is the only one possible, when, in addition to RIEMANN's hypotheses, we accept, from the beginning, that of the mobility of figures, as it would be given by the movement of *Rigid Bodies*.

The problem of RIEMANN-HELMHOLTZ was carefully examined by S. LIE [1842—1899]. He started from the fundamental idea, recognized by KLEIN in HELMHOLTZ's work, *that the congruence of two figures signifies that they are able to be transformed the one into the other, by means of a certain point transformation in space: and that the properties, in virtue of which congruence takes the logical character of equality, depend upon the fact that displacements are given by a group of transformations.*[1]

In this way the problem of RIEMANN-HELMHOLTZ was reduced by LIE to the following form:

Nachr. Bd. XV, p. 193—221 (1868). Wiss. Abhandl., Bd. II, p. 618—639.

The Axioms of Geometry. The Academy, Vol. I, p. 123—181 (1870); Revue des cours scient., T. VII, p. 498—501 (1870).

Über die Axiome der Geometrie. Populäre wissenschaftliche Vorträge, Heft 3, p. 21—54. (Brunswick, 1876). English translation; Mind, Vol. I, p. 301—321. French translation; Revue scientifique de la France et de l'Étranger (2). T. XII, p. 1197—1207 (1877)

Über den Ursprung, Sinn, und Bedeutung der geometrischen Sätze, Wiss. Abh. Bd. II, p. 640—660. English translation; Mind, Vol. II, p. 212—224 (1878).

[1] Cf. Klein: *Vergleichende Betrachtungen über neuere geometrische Forschungen,* (Erlangen, 1872); reprinted in Math. Ann. Bd. XLIII, p. 63—100 (1893). Italian translation by G. FANO, Ann. di Mat. (2), T. XVII, p. 301—343 (1899).

To determine all the continuous groups in space which, in a bounded region, have the property of displacements.

When these properties, which depend upon the free mobility of line and surface elements through a point, are put in a suitable form, there arise *three types of groups*, which characterise the three geometries of EUCLID, of LOBATSCHEWSKY-BOLYAI and of RIEMANN. [1]

Projective Geometry and Non-Euclidean Geometry. Subordination of Metrical Geometry to Projective Geometry.

§ 79. In conclusion, there is an interesting connection between Projective Geometry and the three geometrical systems of EUCLID, LOBATSCHEWSKY-BOLYAI and RIEMANN.

To give an idea of this last method of treating the question, we must remember that Projective Geometry, in the system of G. C. STAUDT [1798—1867], rests simply upon *graphical* notions on the relations between points, lines and planes. Every conception of congruence and movement [and thus of measurement etc.,] is systematically banished. For this reason Projective Geometry, excluding a certain group of postulates, will contain a more restricted number of general properties, which for plane figures are the [*projective*] properties, remaining invariant by projection and section.

However, when we have laid the foundations of Projective Geometry in space, *we can introduce into this system*

---

[1] Cf. LIE: *Theorie der Transformationsgruppen.* Bd. III, p. 437 —543 (Leipzig, 1893). In connection with the same subject, H. POINCARÉ, in his memoir: *Sur les hypothèses fondamentaux de la géométrie* [Bull. de La Soc. Math. de France. T. XV, p. 203—216 (1877)], solved the problem of finding all the hypotheses, which distinguish the fundamental group of plane Euclidean Geometry from the other transformation groups.

the metrical conceptions, as relations between its figures and certain definite (metrical) entities.

Keeping to the case of the Euclidean plane, let us see what *graphical* interpretation can be given to *the fundamental metrical conceptions of parallelism and of perpendicularity.*

To this end we must specially consider *the line at infinity* of the plane, and the *absolute involution* which the set of orthogonal lines of a pencil determine upon it. The double points of such an involution, conjugate imaginaries, are called the *circular points* (at infinity), since they are common to all circles in the plane [PONCELET, 1822 [1]].

On this understanding, *the parallelism* of two lines is expressed graphically *by the property which they possess of meeting in a point on the line at infinity: the perpendicularity* of two lines is expressed graphically *by the property that their points at infinity are conjugate in the absolute involution, that is, form a harmonic range with the circular points.* [CHASLES, 1850.[2]]

Other metrical properties, which can be expressed graphically, are those relative to the size of angles, since every equation

$$F(A, B, C \ldots) = O,$$

between the angles A, B, C, . . ., can be replaced by

$$F \left(\frac{\log a}{2i}, \frac{\log b}{2i}, \frac{\log c}{2i} \ldots \right) = O,$$

in which a, b, c . . . are the *anharmonic ratios* of the pencils formed by the lines bounding the angles and the (imaginary) ines joining the angular points to the circular points. [LA-GUERRE, 1853.[3]]

---

[1] *Traité des propriétés projectives des figures.* 2. Ed., T. I. Nr. 94, p. 48 (Paris, G. Villars, 1865).

[2] *Traité de Géométrie supérieure.* 2. Ed., Nr. 660, p. 425 (Paris, G. Villars, 1880).

[3] *Sur la théorie des foyers.* Nouv. Ann. T. XII, p. 57. Oeuvres de Laguerre. T. II, p. 12—13 (Paris, G. Villars, 1902).

More generally it can be shown that the congruence of any two plane figures can be expressed by a graphical relation between them, the line at infinity, and the absolute involution.[1] Also, since congruence is the foundation of all metrical properties, it follows that the line at infinity and the absolute involution allow all the properties of Euclidean metrical geometry to be subordinated to Projective Geometry. *Thus the metrical properties appear in projective geometry, not as graphical properties of the figures considered in themselves, but as graphical properties with regard to the fundamental metrical entities, made up of the line at infinity and the absolute involution.*

The complete set of fundamental metrical entities is called *the absolute of the plane* (CAYLEY).

All that has been said with regard to the plane can naturally be extended to space. The fundamental metrical entities in space, which allow the metrical properties to be subordinated to the graphical, are the *plane at infinity and a certain polarity (absolute polarity)* on this plane. This polarity is given by the polarity of the sheaf, in which every line corresponds to a plane to which it is perpendicular [cf. § 71]. The fundamental conic of this polarity is *imaginary*, since there are no real lines in the sheaf, which lie on the corresponding perpendicular plane. It can easily be shown that it contains all the pairs of circular points, which belong to the different planes in space, and that it appears as the common section of all spheres. From this property the name of *circle at infinity* is given to this fundamental metrical entity in space.

---

[1] Cf., e. g. F. ENRIQUES, *Lezioni di Geometria proiettiva*, 2a. Ed. p. 177—188 (Bologna, Zanichelli, 1904). There is a German translation of the first edition of this work by H. FLEISCHER (Leipzig, 1903).

§ **80.** The two following questions naturally arise at this stage:

(i) *Can projective geometry be founded upon the Non-Euclidean hypothesis?*

(ii) *If such a foundation is possible, can the metrical properties, as in the Euclidean case, be subordinated to the projective?*

To both these questions the reply is in the affirmative. If RIEMANN's system is valid in space, the foundation of projective geometry does not offer any difficulty, since those graphical properties are immediately verified, which give rise to the ordinary projective geometry, after the *improper entities* are introduced. If the system of LOBATSCHEWSKY-BOLYAI is valid in space, we can also again lay the foundation of the projective geometry, by introducing, with suitable conventions, *improper or ideal points, lines and planes*. This extension will follow the same lines as were taken in the Euclidean case, in completing space with the elements at infinity. It would be sufficient, for this, to consider along with the *proper sheaf* (the set of lines passing through a point), two *improper* sheaves, one formed by all the lines which are parallel to a given line in one direction, the other by all the lines perpendicular to a given plane; also to introduce *improper points*, to be regarded as the *vertices* of these sheaves.

Even if the improper points of a plane cannot in this case, as in the Euclidean, be assigned to a straight line [*the line at infinity*], yet they form a complete region, separated from the region of ordinary points (*proper* points) by a conic [*limiting conic, or conic at infinity*]. This conic is the locus of the *improper* points determined by the pencils of parallel lines.

In space the *improper* points are separated from the *proper* points by a *non-ruled quadric* [*limiting quadric or quadric at infinity*], which is the locus of the improper points determined by sets of parallel lines.

The validity of projective geometry having been established on the Non-Euclidean hypotheses [KLEIN [1]], to obtain the subordination of the metrical geometry to the projective it is sufficient to consider, as in the Euclidean case, the *fundamental metrical entities (the absolute)*, and to interpret the metrical properties of figures as graphical relations between them and these entities. On the plane of LOBATSCHEWSKY-BOLYAI the fundamental metrical entity is the limiting conic, which separates the region of proper points from that of improper points, on the plane of RIEMANN it is an *imaginary conic*, defined by the *absolute polarity* of the plane [cf. p. 144].

In the one case as well as in the other, *the metrical properties of figures are all the graphical properties which remain unaltered in the projective transformations* [2] *leaving the absolute fixed.*

These projective transformations constitute the ∞^3 displacements of the Non-Euclidean plane.

In the Euclidean case the said transformations, (which leave the absolute unaltered), are the ∞^4 transformations of similarity, among which, as a special case, are to be found the ∞^3 displacements.

In space the subordination of the metrical to the pro-

---

[1] The question of the independence of Projective Geometry from the theory of parallels is touched upon lightly by KLEIN in his first memoir: *Über die sogenannte Nicht-Euklidische Geometrie*, Math. Ann. Bd. IV, p. 573—625 (1871). He gives a fuller treatment of the question in Math. Ann. Bd. VI, p. 112—145 (1873). This question is discussed at length in our Appendix IV p. 227.

[2] By the term projective transformation is understood such a transformation as causes a point to correspond to a point, a line to a line, and a point and a line through it, to a point and a line through it.

jective geometry is carried out by means of the limiting quadric (*the absolute of space*). If this is real, we obtain the geometry of LOBATSCHEWSKY-BOLYAI; if it is imaginary, we obtain RIEMANN's elliptic type.

The metrical properties of figures are therefore the graphical properties of space in relation to its absolute; that is, the graphical properties which remain unaltered in all the projective transformations which leave the absolute of space fixed.

§ 81. How will the ideas of distance and of angle be expressed with reference to the absolute?

Take a system of homogeneous coordinates (x_1, x_2, x_3) on the projective plane. By their means the straight line is represented by a linear equation, and the equation of the absolute takes the form:

$$\Omega_{xx} = \Sigma a_{ij}\, x_i\, x_j = 0.$$

Then the distance between two points $X\,(x_1, x_2, x_3)$, $Y\,(y_1, y_2, y_3)$ is expressed, omitting a constant factor, *by the logarithm of the anharmonic ratio of the range consisting of X, Y, and the points M, N, in which the line $X\,Y$ meets the absolute.*

If we then put

$$\Omega_{xy} = \Sigma a_{ij}\, x_i\, y_j,$$

and remember, from analytical geometry, that the anharmonic ratio of the four points X, Y, M, N is given by

$$\frac{\Omega_{xy} + \sqrt{\Omega_{xy}^2 - \Omega_{xx}\,\Omega_{yy}}}{\Omega_{xy} - \sqrt{\Omega_{xy}^2 - \Omega_{xx}\,\Omega_{yy}}},$$

the expression for the distance D_{xy} will be:—

$$(1) \qquad D_{xy} = \frac{k}{2} \log \frac{\Omega_{xy} + \sqrt{\Omega_{xy}^2 - \Omega_{xx}\,\Omega_{yy}}}{\Omega_{xy} - \sqrt{\Omega_{xy}^2 - \Omega_{xx}\,\Omega_{yy}}}.$$

Introducing the inverse circular and hyperbolic functions,

$$(2) \quad \begin{cases} D_{xy} = ik \, \cos^{-1} \dfrac{\Omega_{xy}}{\sqrt{\Omega_{xx}\Omega_{yy}}} \\[2em] D_{xy} = k \, \cosh^{-1} \dfrac{\Omega_{xy}}{\sqrt{\Omega_{xx}\Omega_{yy}}}, \end{cases}$$

$$(3) \quad \begin{cases} D_{xy} = ik \, \sin^{-1} \dfrac{\sqrt{\Omega_{xx}\Omega_{yy} - \Omega_{xy}^2}}{\sqrt{\Omega_{xx}\Omega_{yy}}} \\[2em] D_{xy} = k \, \sinh^{-1} \dfrac{\sqrt{\Omega_{xy}^2 - \Omega_{xx}\Omega_{yy}}}{\sqrt{\Omega_{xx}\Omega_{yy}}}. \end{cases}$$

The constant k, which appears in these formulæ, is connected with RIEMANN's Curvature K by the equation

$$K = -\frac{1}{k^2}.$$

Similar considerations lead to the projective interpretation of the conception of angle. *The angle between two lines is proportional to the logarithm of the anharmonic ratio of the pencil which they form with the tangents from their point of intersection to the absolute.*

If we wish the measure of the complete pencil to be 2π, as in the ordinary measurement, we must take the fraction $1 : 2i$ as the constant multiplier. Then to express analytically the angle between two lines u (u_1, u_2, u_3), v (v_1, v_2, v_3), we put

$$\Psi_{uu} = \Sigma \, b_{ij} \, u_i \, u_j.$$

If b_{ij} is the cofactor of the element a_{ij} in the discriminant of Ω_{xx}, the tangential equation of the absolute is given by

$$\Psi_{uu} = 0,$$

and the angle between the two lines by the following formulæ:—

$$(1') \qquad \measuredangle u, v = \frac{1}{2i} \log \frac{\Psi_{uv} + \sqrt{\Psi_{uv}^2 - \Psi_{uu}\Psi_{vv}}}{\Psi_{uv} - \sqrt{\Psi_{uv}^2 - \Psi_{uu}\Psi_{vv}}},$$

$$(2')\begin{cases} \measuredangle\, u, v = \cos^{-1}\dfrac{\Psi_{uv}}{\sqrt{\Psi_{uu}\,\Psi_{vv}}} \\[2em] \measuredangle\, u, v = \dfrac{1}{i}\cosh^{-1}\dfrac{\Psi_{uv}}{\sqrt{\Psi_{uu}\,\Psi_{vv}}}, \end{cases}$$

$$(3')\begin{cases} \measuredangle\, u, v = \sin^{-1}\dfrac{\sqrt{\Psi_{uu}\,\Psi_{vv} - \Psi_{uv}^2}}{\sqrt{\Psi_{uu}\,\Psi_{vv}}} \\[2em] \measuredangle\, u, v = \dfrac{1}{i}\sinh^{-1}\dfrac{\sqrt{\Psi_{uv}^2 - \Psi_{uu}\,\Psi_{vv}}}{\sqrt{\Psi_{uu}\,\Psi_{vv}}}. \end{cases}$$

Similar expressions hold for the distance between two points and the angle between two planes, in the geometry of space. We need only suppose that

$$\Omega_{xx} = 0,\ \Psi_{uu} = 0,$$

represent the equations (in point and tangential coordinates) of the absolute of space, instead of the absolute of the plane.

According as $\Omega_{xx} = O$ is the equation of a real quadric, without generating lines, or of an imaginary quadric, the formulæ will refer to the geometry of LOBATSCHEWKY-BOLYAI, or that of RIEMANN.[1]

§ 82. The preceding formulæ, concerning the angles between two lines or planes, contain those of ordinary geometry as a special case. Indeed if, for simplicity, we take the case of the plane, and the system of orthogonal axes, the tangential equation of the Euclidean absolute (*the circular points*, § 79) is

$$u_1{}^2 + u_2{}^2 = O.$$

The formula $(2')$, when we insert

$$\Psi_{uu} = u_1{}^2 + u_2{}^2,\ \Psi_{vv} = v_1{}^2 + v_2{}^2,\ \Psi_{uv} = u_1 v_1 + u_2 v_2,$$

becomes

---

[1] For a full discussion of the subject of this and the preceding sections, see CLEBSCH-LINDEMANN, *Vorlesungen über Geometrie*, Bd. II. Th. I, p. 461—et seq. (Leipzig, 1891).

$$\measuredangle\ u,\ v = \cos^{-1} \frac{u_1 v_1 + u_2 v_2}{\sqrt{u_1{}^2 + u_2{}^2} \cdot \sqrt{v_1{}^2 + v_2{}^2}},$$

from which we have

$$\cos(u,\ v) = \frac{u_1 v_1 + u_2 v_2}{\sqrt{u_1{}^2 + u_2{}^2} \cdot \sqrt{v_1{}^2 + v_2{}^2}}.$$

But the direction cosines of the line u (u_1, u_2, u_3) are

$$\cos(u,\ x) = \frac{u_1}{\sqrt{u_1{}^2 + u_2{}^2}}, \qquad \cos(u\,y) = \frac{u_2}{\sqrt{u_1{}^2 + u_2{}^2}},$$

so that this equation can be written

$$\cos(u,v) = l_1 l_2 + m_1 m_2$$

the ordinary expression for the angle between the two lines $(l_1 m_1)$ and $(l_2 m_2)$.

For the distance between two points the argument does not proceed so simply, when the absolute degenerates into the circular points. Indeed the points M, N, where the line XY intersects the absolute, coincide in the point at infinity on this line, and the formula (1) gives in every case:

$$D_{xy} = \frac{k}{2} \log(M_\infty N_\infty XY) = \frac{k}{2} \log 1 = 0.$$

However, by a simple artifice we can obtain the ordinary formula for the distance as the limiting case of formula (3).

To do this more easily, let us suppose the equations of the absolute (not degenerate), in point and line coordinates, reduced to the form:

$$\Omega_{xx} = \epsilon x_1{}^2 + \epsilon x_2{}^2 + x_3{}^2 = 0,$$
$$\Psi_{uu} = u_1{}^2 + u_2{}^2 + \epsilon u_3{}^2 = 0.$$

Then, putting

$$\Delta = \frac{\sqrt{\epsilon(x_1 y_2 - y_1 x_2)^2 + (x_1 y_3 - y_1 x_3)^2 + (x_2 y_3 - x_3 y_2)^2}}{\sqrt{\epsilon x_1{}^2 + \epsilon x_2{}^2 + x_3{}^2} \cdot \sqrt{\epsilon y_1{}^2 + \epsilon y_2{}^2 + y_3{}^2}},$$

equation (3) of the preceding section gives

$$D_{xy} = ik \sin^{-1} \sqrt{\epsilon} \, \Delta.$$

Let ϵ be infinitesimal. Omitting terms of a higher order, we can substitute $\sqrt{\epsilon}\Delta$ for $\sin^{-1}\sqrt{\epsilon}\Delta$ in this formula. If we now choose k^2 infinitely large, so that the product $ik\sqrt{\epsilon}$ remains finite and equal to unity for every value of ϵ, the said formula becomes

$$D_{xy} = \frac{\sqrt{\epsilon(x_1 y_2 - y_1 x_2)^2 + (x_1 y_3 - y_1 x_3)^2 + (x_2 y_3 - y_2 x_3)^2}}{\sqrt{\epsilon x_1^2 + \epsilon x_2^2 + x_3^2} \; \sqrt{\epsilon y_1^2 + \epsilon y_2^2 + y_3^2}}$$

Let ϵ now tend to the limit zero. The tangential equation of the absolute becomes

$$u_1^2 + u_2^2 = 0;$$

and the conic degenerates into two imaginary conjugate points on the line $u_3 = o$. The formula for the distance, on putting

$$X_i = \frac{x_i}{x_3}, \; Y_i = \frac{y_i}{y_3},$$

takes the form

$$D_{xy} = \sqrt{(X_1 - Y_1)^2 + (X_2 - Y_2)^2},$$

which is the ordinary Euclidean formula. We have thus obtained the required result.

We note that to obtain the special Euclidean case from the general formula for the distance, we must let k^2 tend to infinity. Since RIEMANN's curvature is given by $-\frac{1}{k^2}$, this affords a confirmation of the fact that RIEMANN's curvature is zero in Euclidean space.

§ 83. The properties of plane figures with respect to a conic, and those of space with respect to a quadric, together constitute projective metrical geometry. This was first studied by CAYLEY,[1] apart from its connection with the Non-Euclid-

---

[1] *Sixth Memoir upon Quantics.* Phil. Trans. Vol. CXLIX, p. 61 —90 (1859). Also *Collected Works*, Vol. II, p. 561—592.

ean geometries. These last relations were discovered and explained some years later by F. KLEIN. [1]

To KLEIN is also due a widely used nomenclature for the projective metrical geometries. He gives the name *hyperbolic geometry* to CAYLEY's geometry, when the absolute is real and not degenerate: *elliptic geometry*, to that in which the absolute is imaginary and not degenerate: *parabolic geometry*, to the limiting case of these two. Thus, in the remaining articles, we can use this nomenclature to describe the three geometrical systems of LOBATSCHEWSKY-BOLYAI, of RIEMANN (elliptic type), and of EUCLID.

Representation of the Geometry of Lobatschewsky-Bolyai on the Euclidean Plane.

§ 84. To the projective interpretation of the Non-Euclidean measurements, of which we have just spoken, may be added an interesting representation which can be given of the *Hyperbolic Geometry* on the Euclidean plane. To obtain it, we take on the plane a real, not degenerate, conic: e. g. a circle. Then we make the following definitions, relative to this circle:

Plane = region of points within the circle.

Point = point inside the circle.

Straight line = chord of the circle.

We can now easily verify that the postulate that a straight line is determined by two points, and the postulates regarding the properties of straight lines and angles, can be expressed as relations, which are always valid, when the above interpretations are given to these terms.

But in the further development of this geometry we add

---

[1] Cf. *Über die sogenannte Nicht-Euklidische Geometrie.* Math. Ann. Bd. IV, p. 573—625 (1871).

to these the postulates of congruence, contained in the following *principle of displacement.*

If we are given two points A, A' on the plane, and the straight lines a, a', respectively passing through them, there are four methods of superposing the plane on itself, so that A and a coincide respectively with A' and a'. More precisely: *one* method of superposition is defined by taking as corresponding to each other, one ray of a and one ray of a', one section of the plane bounded by a and one section bounded by a'. Two of these displacements are *direct congruences* and two *converse congruences.*

With the preceding interpretations of the entities, *point, line* and *plane*, the principle here expressed is translated into the following proposition:

If a conic (e. g., a circle) is given in a plane, and two internal points A, A' are taken, as also two chords a, a', respectively passing through them, there are four projective transformations of the plane, which change into itself the space within the conic, and which make A and a correspond respectively to A' and a'.

To fix *one* of them, it is sufficient to make sure that a given extremity of a corresponds to a given extremity of a', and that to one section of the plane bounded by a, corresponds a definite section of the plane bounded by a'. Of these four transformations, two determine on the conic *a projective correspondence in the same sense*, and two *a projective correspondence in the opposite sense.*

§ **85.** We shall prove this proposition, taking for simplicity two distinct conics τ, τ', in the same plane or otherwise.

Let M, N be the extremities of the chord a [cf. Fig. 58]. Also M', N' those of a' [cf. Fig. 59].

Let P, P' be the poles of a, a' with respect to the two conics.

On this understanding, the line PA intersects the conic τ in two real and distinct points R, S: also the line $P'A'$ intersects the conic τ′ in two real and distinct points R', S'.

A projective transformation which changes τ into τ′, the line a into a', and the point A into A', will make the point P correspond to P', and the line PA to the line $P'A'$.

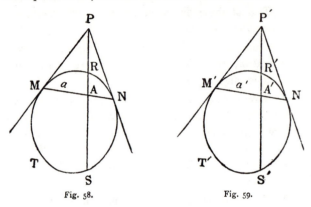

Fig. 58.　　　　　　Fig. 59.

Thus this transformation determines a projective correspondence between the points of the two conics, in which the pair of points M', N' corresponds to the pair of points M, N: and the pair of points R', S' to R, S.

Vice versa, a projective transformation between the two conics, which enjoys this property, is associated with a projective transformation of the two planes, such as is here described.[1]

But if we consider the two conics τ, τ′, we see that to

---

[1] For this proof, and the theorems of Projective Geometry upon which it is founded, see Chapter X, p. 251—253 of the work of ENRIQUES referred to on p. 156.

the points of the range *MNRS* on τ may be made to correspond the points of any one of the following ranges on τ′:

$$M'N'R'S'$$
$$N'M'S'R'$$
$$M'N'S'R'$$
$$N'M'R'S'.$$

In this way we prove the existence of the four projective transformations of which we have spoken in the proposition just enunciated.

If we suppose that the two conics coincide, we do not need to change the preceding argument in any way. We add, however, that of the four transformations only one makes the segment *AM* correspond to the segment *A′M′*, if at the same time the shaded parts of the figure correspond to each other.

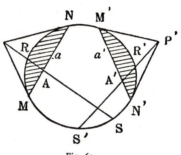

Fig. 60.

Further the two transformations defined by the ranges

$$\left(\begin{array}{c} MNRS \\ M'N'R'S' \end{array} \right), \quad \left(\begin{array}{c} MNRS \\ N'M'S'R' \end{array} \right)$$

determine *projections in the same sense*, while the other two, defined by the ranges:

$$\left(\begin{array}{c} MNRS \\ M'N'S'R' \end{array} \right) \quad \left(\begin{array}{c} MNRS \\ N'M'R'S' \end{array} \right)$$

determine *projections in the opposite sense*.

§ 86. With these remarks, we now return to complete the definitions of § 84, relative to a circle given on the plane.

Plane = region of points within the circle.

Point = point within the circle.

Straight Line = chord of the circle.

Displacements = projective transformations of the plane which change the space within the circle into itself.

Semi-Revolutions = homographic transformations of the circle.

Congruent Figures = figures which can be transformed the one into the other by means of the projective transformations named above.

The preceding arguments permit us to affirm at once that all the propositions of elementary plane geometry, associated with the concepts straight line, angle and congruence, can be readily translated into properties relative to the system of points inside the circle, which we denote by (S). In particular let us see what corresponds in (S) to two perpendicular lines in the ordinary plane.

To this end we note that if r, s are two perpendicular lines, a semi-revolution of the plane about s will superpose r upon itself, exchanging, however, the two rays in which it is divided by s.

According to the above definitions, a semi-revolution in (S) is a homographic transformation, which has for *axis* a chord s of the circle and for *centre* the pole of the chord. The lines which are unchanged in this transformation, in addition to s, are the lines passing through its centre. *Thus in the system (S) we must call two lines perpendicular, when they are conjugate with respect to the fundamental circle.*

We could easily verify in (S) all the propositions on perpendicular lines. In particular, that if we draw the (imaginary) tangents to the fundamental circle from the common point of two conjugate chords in (S), these tangents form a harmonic pencil with the perpendicular lines [cf. p. 155].[1]

---

[1] This representation of the Non-Euclidean plane has been

§ 87. Let us now see how the *distance* between two points can be expressed in this conventional measurement, which is being taken for the interior of the circle.

To this end we introduce a system of orthogonal coordinates (x, y), with origin at the centre of the circle.

The *distance* between two points A (x, y), B (x', y') in the plane with which we are dealing cannot be represented by the usual formula

$$\sqrt{(x-x')^2 + (y-y')^2},$$

since it is *not invariant* for the projective transformations which we have called displacements. The *distance* must be a function of the coordinates, invariant for the said transformations, which for points on the straight line possesses the distributive property given by the formula

dist. (AB) = dist. (AC) + dist. (CB).

Now the anharmonic ratio of the four points A, B, M, N, where M, N are the extremities of the chord AB, is a relation between the coordinates (x, y), (x', y') of AB, remaining invariant for all projective transformations which leave the fundamental circle fixed. The most general expression, possessing this invariant property, will be an arbitrary function of this anharmonic ratio.

If we remember that the said function must be distributive in the sense above indicated, we must assume that, except for a multiplier, it is equal to the logarithm of the anharmonic ratio,

$$(ABMN) = \frac{AM}{BM} : \frac{AN}{BN}.$$

We shall thus have

distance $(AB) = \frac{k}{2} \log (ABMN)$.

---

employed by GROSSMANN in carrying out a number of the constructions of Non-Euclidean Geometry. Cf. Appendix, III, p. 225.

In a similar way we proceed to find the proper expression for the angle between two straight lines. In this case we must notice that if we wish the right angle to be expressed by $\frac{\pi}{2}$, we must take as constant multiplier of the logarithm the factor $1 : 2i$.

Then we shall have for the angle between a and b,

$$\sphericalangle\, a, b = \frac{1}{2i} \log\, (abmn),$$

where m, n are the conjugate imaginary tangents from the vertex of the angle to the circle, and $(a\,b\,m\,n)$ is the anharmonic ratio of the four lines a, b, m and n, expressed analytically by

$$\frac{\sin\, (a\,m)}{\sin\, (b\,m)} : \frac{\sin\, (a\,n)}{\sin\, (b\,n)}.$$

§ 88. A glance at what was said above on the subordination of the metrical to the projective geometry (§ 81) will show clearly that the preceding formulæ, regarding the *distance* and *angle*, agree with those which we would have in the Non-Euclidean plane, if the absolute were a circle. This would be sufficient to suggest that the geometry of the system (S) gives a concrete representation of the geometry of LOBATSCHEWSKY-BOLYAI. However, as we wish to discuss this point more fully, let us see how the definition and property of parallels are translated in (S).

Let $r\, (u_1, u_2, u_3)$ and $r'\, (v_1, v_2, v_3)$ be two different chords of the fundamental circle.

Let the circle be referred to an orthogonal Cartesian set of axes, with the centre for origin, and let us take the radius as unit of length.

Then we have

$$x^2 + y^2 - 1 = 0,$$
$$u^2 + v^2 - 1 = 0,$$

for the point and line equation of the circle.

Making these equations homogeneous, we obtain

$$x_1{}^2 + x_2{}^2 - x_3{}^2 = 0,$$
$$u_1{}^2 + u_2{}^2 - u_3{}^2 = 0.$$

The angle $\measuredangle r, r'$ between the two straight lines r and r can be calculated by means of the formula ($3'$) of § 81, if we put

$$\Psi_{uu} = u_1{}^2 + u_2{}^2 - u_3{}^2,$$
$$\Psi_{vv} = v_1{}^2 + v_2{}^2 - v_3{}^2,$$
$$\Psi_{uv} = u_1 v_1 + u_2 v_2 - u_3 v_3.$$

We thus obtain

$$\sin \measuredangle r, r' = \frac{\sqrt{(u_1 v_2 - v_1 u_2)^2 - (u_2 v_3 - v_3 u_2)^2 - (u_3 v_1 - v_1 u_3)^2}}{\sqrt{(u_1{}^2 + u_2{}^2 - u_3{}^2)(v_1{}^2 + v_2{}^2 - v_3{}^2)}}.$$

But the lines r, r' are given by

$$x_1 u_1 + x_2 u_2 + x_3 u_3 = 0,$$
$$x_1 v_1 + x_2 v_2 + x_3 v_3 = 0;$$

and they meet in the point,

$$x_1 = u_2 v_3 - u_3 v_2,$$
$$x_2 = u_3 v_1 - u_1 v_3,$$
$$x_3 = u_1 v_2 - u_2 v_1.$$

Thus the preceding expression for this angle takes the form

$$(4) \quad \sin \measuredangle r, r' = \frac{\sqrt{(x_3{}^2 - x_1{}^2 - x_2{}^2)}}{\sqrt{(u_1{}^2 + u_2{}^2 - u_3{}^2)(v_1{}^2 + v_2{}^2 - v_3{}^2)}}.$$

From this it is evident that the necessary and sufficient condition that the angle be zero is that the numerator of this fraction should vanish.

Now if this numerator is zero, the point (x_1, x_2, x_3), in which the chords intersect, must lie on the circumference of the fundamental circle, and vice versa (Fig. 61).

Therefore in our interpretation of the geometrical propositions by means of the system (S), we must call two chords parallel, when they meet in a point on the circumference of the fundamental circle, since the angle between those two chords is zero.

Since there are two chords through any point within a circle which join this point to the ends of any given chord, the fundamental proposition of hyperbolic geometry will be verified for the system (S).

§ 89. We proceed to find for the system (S) the formula regarding the angle of parallelism. To do this we first calculate the angle OMN, between the axis of y and the line MN, joining a point M on the axis of y to the extremity of the axis of x (Fig. 62).

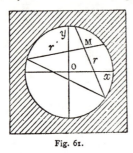
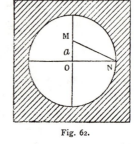

Fig. 61.   Fig. 62.

Denoting by a the ordinary distance of the two points M and O, the homogeneous coordinates of the line MN and the line OM are, respectively $(a, 1, -a)$, $(1, 0, 0)$ and the coordinates of their common point are $(0, a, 1)$.

Then from (4) of the preceding article,

$$\sin \angle OMN = \sqrt{1-a^2}.$$

On the other hand, the distance, according to our convention, between the two points O and M is given by (2) of § 81 as

$$OM = k \cosh^{-1} \frac{1}{\sqrt{1-a^2}}.$$

Thus

$$\cosh \frac{OM}{k} = \frac{1}{\sqrt{1-a^2}}.$$

Comparing these two results, we have

$$\cosh \frac{OM}{k} = \frac{1}{\sin \angle OMN},$$

a relation which agrees with that given by TAURINUS, LO-BATSCHEWSKY and BOLYAI for the angle of parallelism [cf. p. 90].

§ 90. We proceed, finally, to see how the distance between two neighbouring points (*the element of distance*) is expressed in the system (S), so that we may be able to compare this representation of the hyperbolic geometry with that given by BELTRAMI [cf. § 69].

Let (x, y), $(x + dx, y + dy)$ be two neighbouring points. Their distance ds is calculated by means of (2) of § 81 if we substitute:

$$\Omega_{xx} = x^2 + y^2 - 1,$$
$$\Omega_{yy} = (x + dx)^2 + (y + dy)^2 - 1,$$
$$\Omega_{xy} = x(x + dx) + y(y + dy) - 1.$$

Since the angle is small, we may substitute the sine for the angle, and we have

$$ds^2 = k^2 \frac{(dx^2 + dy^2)(1 - x^2 - y^2) + (x\,dx + y\,dy)^2}{(x^2 + y^2 - 1)((x + dx)^2 + (y + dy)^2 - 1))}.$$

Thus, omitting terms higher than the second order, we have

$$ds^2 = k^2 \frac{(dx^2 + dy^2)(1 - x^2 - y^2) + (x\,dx + y\,dy)^2}{(1 - x^2 - y^2)^2}$$

or

$$(5) \quad ds^2 = k^2 \frac{(1 - y^2)\,dx^2 + 2xy\,dx\,dy + (1 - x^2)\,dy^2}{(1 - x^2 - y^2)^2}.$$

Now we recall that BELTRAMI, in 1868, interpreted the geometry of LOBATSCHEWSKY-BOLYAI by that on the surfaces of constant negative curvature. The study of the geometry on such surfaces depends upon the use of a system of coordinates on the surface, and the law according to which the *element of distance* (ds) is measured. The choice of a suitable system (u, v) enabled BELTRAMI to put the square of ds in this form:

$$k^2 \; \frac{(1 - v^2)\, du^2 + 2\, u v\, du\, dv + (1 - u^2)\, dv^2}{(1 - u^2 - v^2)^2},$$

where the constant k^2 is the reciprocal, with its sign changed, of the curvature of the surface.[1]

In studying the properties of these surfaces and in making a comparison between them and the metrical results of the geometry of LOBATSCHEWSKY-BOLYAI, BELTRAMI in his classical memoir, quoted on p. 138, employed the following artifice:

He represented the points of the surface on an auxiliary plane, such that the point (u, v) of the surface corresponded to the point on the plane whose Cartesian coordinates (x, y) were (u, v). The points on the surface were then represented by points inside the circle

$$x^2 + y^2 - 1 = 0;$$

the points at infinity on the surface by points on the circumference of the circle: its geodesics by chords: parallel geodesics by chords meeting in a point on the circumference of the said circle. Then the expression for $(ds)^2$ took the same form as that given in (5), which states the form to be used for the element of distance in the system (S).

It follows that, by his representation of the surfaces of constant negative curvature on a plane, BELTRAMI was led to one of the projective metrical geometries of CAYLEY, and precisely to the metrical geometry relative to a fundamental circle, given above in §§ 80, 81.

[1] *Risoluzione del problema di riportare i punti di una superficie sopra un piano in modo che le linee geodetiche vengano rappresentate da linee rette.* Ann. di Mat. T. VII, p. 185—204 (1866). Also *Opere Matematiche.* T. I, p. 262—280 (Milan, 1902).

§ 91. The representation of plane hyperbolical geometry on the Euclidean plane is capable of being extended to
the case of solid geometry. To represent the solid geometry
of LOBATSCHEWSKY-BOLYAI in ordinary space we need only
adopt the following definitions for the latter:

Space = Region of points inside a sphere.

Point = Point inside the sphere.

Straight Line = Chord of the sphere.

Plane = Points of a plane of section which are inside
the sphere.

Displacements = Projective transformations of space,
which change the region of the points inside the
sphere into itself, etc.

With this 'Dictionary' the propositions of hyperbolic
solid geometry can be translated into corresponding properties of the Euclidean space, relative to the system of points
inside the sphere.[1]

Representation of Riemann's Elliptic Geometry in Euclidean Space.

§ 92. So far as regards plane geometry, we have already
remarked [pp. 142—3] that the geometry of the ordinary
sheaf of lines gives a concrete interpretation of the elliptical
system of RIEMANN. Therefore, if we cut the sheaf by an
ordinary plane, completed by the line at infinity, we obtain
a representation on the Euclidean plane of the said RIE-
MANN's plane.

---

[1] BELTRAMI considers the interpretation of Non-Euclidean Solid
Geometry, and, in general, of the geometries of manifolds of
higher order in space of constant curvature, in his memoir: *Teoria
fondamentale degli spazii di curvatura costante.* Ann. di Mat. (2),
T. II, p. 232—255 (1868). *Opere Mat.* T. I, p. 406—429 (Milan,
1902).

If we wish a representation of the elliptic space in the Euclidean space, we need only assume in this a single-valued polarity, *to which corresponds an imaginary quadric, not degenerate.* We must then take, with respect to this quadric, a system of definitions analogous to those indicated above in the hyperbolic case. We do not pursue this point further, as it offers no fresh difficulty.

However we remark that in this representation *all the points of the Euclidean space, including the points on the plane at infinity, would have a one-one correspondence with the points of Riemann's space.*

Foundation of Geometry upon Descriptive Properties.

§ 93. The principles explained in the preceding sections lead to a new order of ideas in which the descriptive properties appear as the first foundations of geometry, instead of congruence and displacement, of which RIEMANN and HELMHOLTZ availed themselves. We note that, if we do not wish to introduce at the beginning any hypothesis on the intersection of coplanar straight lines, we must start from a suitable system of postulates, valid in a *bounded region* of space, and that we must complete the initial region later by means of *improper points, lines and planes* [cf. p. 157].[1]

When projective geometry has been developed, the metrical properties can be introduced into space, by adding to the initial postulates those referring to displacement or

---

[1] For such developments, cf. KLEIN, loc. cit. p. 158: PASCH, *Vorlesungen über neuere Geometrie,* (Leipzig, 1882); SCHUR, *Über die Einführung der sogenannten idealen Elemente in die projective Geometrie,* Math. Ann. Bd. XXXIX, p. 113—124 (1891): BONOLA, *Sulla introduzione degli elementi improprii in geometria proiettiva,* Giornale di Mat. T. XXXVIII, p. 105—116 (1900).

congruence. By so doing we find that a certain polarity of space, allied to the metrical conceptions, becomes transformed into itself by all displacements. Then it is shown that the fundamental quadric of this polarity can only be:

   a) *A real, non-ruled quadric;*
   b) *An imaginary quadric* (with real equation);
   c) *A degenerate quadric.*

Thus the *three geometrical systems,* which RIEMANN and HELMHOLTZ reached from the conception of the element of distance, are to be found also in this way.[1]

The Impossibility of proving Euclid's Postulate.

§ 94. Before we bring to a close this historical treatment of our subject it seems advisable to say a few words on the impossibility of demonstrating Euclid's Postulate.

The very fact that the innumerable attempts made to obtain a proof did not lead to the wished-for result, would suggest the thought that its *demonstration is impossible.* Indeed our geometrical instinct seems to afford us evidence that a proposition, seemingly so simple, if it is provable, ought to be proved by an argument of equal simplicity. But such considerations cannot be held to afford a *proof* of the impossibility in question.

If we put EUCLID's Postulate aside, following the developments of GAUSS, LOBATSCHEWSKY and BOLYAI, we can construct a geometrical system in which no contradictions are met. This seems to prove the logical possibility of the Non-Euclidean hypothesis, and that EUCLID's Postulate is *independent* of the first principles of geometry and therefore *cannot be demonstrated.* However the fact that contradictions

---

[1] For the proof of this result see BONOLA, *Determinazione per via geometrica dei tre tipi de spazio; iperbolico, parabolico, ellittico.* Rend. Circ. Mat. Palermo, T. XV, p. 56—65 (1901).

have not been met is not sufficient to prove this; we must be certain that, proceeding on the same lines, such contradictions could never be met. This conviction can be gained with absolute certainty from the consideration of the formulæ of Non-Euclidean geometry. If we take the system of all the sets of three numbers (x, y, z), and agree to consider each set as an *analytical point*, we can define the *distance* between two such analytical points by the formulæ of the said Non-Euclidean Trigonometry. In this way we construct an analytical system, which offers a conventional interpretation of the Non-Euclidean geometry, and thus demonstrates its logical possibility.

In this sense the formulae of the Non-Euclidean Trigonometry of Lobatschewsky-Bolyai give the proof of the independence of Euclid's Postulate from the first principles of geometry (regarding the straight line, the plane and congruence).

We can seek a *geometrical proof* of the said independence, on the lines of the later developments of which we have given an account. For this it is necessary to start from the principle that the conceptions, derived from our intuition, independently of the correspondence which they find in the external world, are a priori *logically possible*; and that thus the Euclidean geometry is logically possible and every set of deductions founded upon it.

But the interpretation which the Non-Euclidean plane hyperbolic geometry finds in the geometry on the surfaces of constant negative curvature, offers, *up to a certain point*, a first proof of the impossibility of demonstrating the Euclidean postulate. To put the matter in more exact terms: *by this means it is established that the said postulate cannot be demonstrated on the foundation of the first principles of geometry, held valid in a bounded region of the plane.* In fact, every contradiction, which would arise from the other postulate, would be translated into a contradiction in the geometry on the surfaces of constant negative curvature.

However, since the comparison between the hyperbolic plane and the surfaces of constant negative curvature, exists, as we have seen, only for *bounded regions*, we have not thus excluded the possibility that the Euclidean postulate might be proved for the *complete plane*.

To remove this uncertainty, it would be necessary to refer to the abstract *manifold* of constant curvature, since no concrete surface exists in ordinary space, in which the *complete* hyperbolic geometry holds [cf. § 73].

But, even so, the impossibility of proving Euclid's Postulate would have been shown only for *plane geometry*. There would still remain the question of the possibility of proving it by means of the *considerations of solid geometry*.

The foundation of geometry, on RIEMANN's principles, whereby the ideas of the geometry on a surface are extended to a three-dimensional region, gives the complete proof of the impossibility of this demonstration. This proof depends on the existence of *a Non-Euclidean analytical system*. Thus we are brought to another analytical proof. The same remark applies also to the investigations of HELMHOLTZ and LIE, though it might be argued that the latter also offer a geometrical proof, from the existence of *transformation groups of the Euclidean space, similar to the groups of displacements of the Non-Euclidean geometry*. Of course, it must be understood that we here consider geometry in its fullest sense.

But the *proof of the impossibility of demonstrating Euclid's Postulate, which is based upon the projective measurements of Cayley*, is simpler and easier to follow geometrically.

This proof depends upon the representation of the Non-Euclidean geometry by the conventional measurement relative to a circle or to a sphere, an interpretation which we have developed at length in the case of the plane [§§ 84 —92].

Further the proof of the logical possibility of RIEMANN's *elliptic hypothesis* can be just as easily derived from these projective measurements. For the plane, the interpretation which we have given of it as the geometry of the sheaf will be sufficient [§ 71].[1]

---

[1] Another neat and simple proof of the independence of the Fifth Postulate is to be found in the representation of the Non-Euclidean plane, employed by KLEIN and POINCARÉ. In this the points of the Non-Euclidean plane appear as points of the upper portion of the Euclidean plane, and the straight lines of the Non-Euclidean plane as semicircles, perpendicular to the straight boundary of this halfplane; etc. The Elliptic Geometry can be represented in a similar way; and the Hyperbolic and Elliptic Solid Geometries can also be brought into correspondence with the Euclidean Space. An account of these representations is to be found in WEBER und WELLSTEIN's *Encyklopädie der Elementar-Mathematik*, Bd. II § 9—11, p. 39—81 (Leipzig, 1905) and in Chapter II of the *Nicht-Euklidische Geometrie* by H. LIEBMANN (Sammlung SCHUBERT, 49, Leipzig, 1905).

In Appendix V of this volume a similar argument is given, based upon the discussion in WEBER-WELLSTEIN's volume. Points upon the Non-Euclidean plane are represented by pairs of points inverse to a fixed circle on the Euclidean plane; and straight lines upon the one, are circles orthogonal to the fixed circle on the other.

Appendix I.

The Fundamental Principles of Statics and Euclid's Postulate.

On the Principle of the Lever.

§ 1. To demonstrate the Principle of the Lever, ARCHI-MEDES [287—212] avails himself of several hypotheses, some expressed and others implied. Among the hypotheses passed over in silence, in addition to that which we would now call *the hypothesis of increased constraint*[1], there is one which definitely concerns the equilibrium of the lever, and can be expressed as follows:

When a lever is suspended from its middle point, it is in equilibrium, if a weight $2P$ is applied at one end, and at the other another lever is hung by its middle point, each of its ends supporting a weight P.[2]

We shall not discuss the various criticisms upon ARCHI-MEDES' use of this hypothesis, nor the different attempts made to prove it.[3] In this connection we shall refer only to the

---

[1] This hypothesis can be enunciated as follows: *If several bodies, subjected to constraints, are in equilibrium under the action of given forces, they will still be in equilibrium, if new constraints are added to those already in existence.* Cf., for example, J. ANDRADE, *Leçons de Mécanique Physique*, p. 59 (Paris, 1898).

[2] Cf. *Archimedis opera omnia:* critical edition by J. L. HEIBERG; Bd. II, p. 142 et seq. (Leipzig, 1881).

[3] Cf., for example, E. MACH, *Die Mechanik in ihrer Ent-*

arguments of LAGRANGE, since these will show, clearly and simply, the important link between this hypothesis and the Parallel Postulate.

§ 2. Let ABD be an isosceles triangle ($AD = BD$), from whose angular points A and B are suspended two (cf. Fig. 63) equal weights P, while a weight equal to $2P$ is suspended from D.

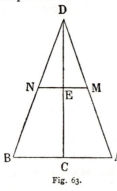

Fig. 63.

This triangle will be in equilibrium about the straight line MN, joining the middle points of the equal sides, since each of these sides may be regarded as a lever from whose extremities *equal* weights are hung.

But the equilibrium of the figure will also be secured, if the triangle rests upon a line passing through the vertex D and the middle point C of the side AB. Therefore, if E is the common point of CD and MN, the triangle will be in equilibrium, when suspended from E.

'Or', continues LAGRANGE, 'comme l'axe [MN] passe par le milieu des deux côtés du triangle, il passera aussi nécessairement par le milieu de la droite menée du sommet du triangle au milieu [C] de sa base; donc le levier transversal [CD] aura le point d'appui [E] dans le milieu et devra, par conséquent, être chargé également aux bouts [C, D]: donc la charge que supporte le point d'appui du levier, qui fait la base du triangle, et qui est chargé, à ses

wickelung, (3. Aufl., Leipzig, 1897); English translation by T. J. Mc-CORMACK (Open Court Publishing Co. Chicago, 1902). Also, for the different hypotheses from which the proof of the principle of the lever, can be obtained, see P. DUHEM, *Les origines de la statique*, (Paris, 1905), especially Appendix C, *Sur les divers axiomes d'où se peut déduire la théorie du levier*.

deux extrémités de poids égaux, sera égale au poids double du sommet et, par conséquent, égale à la somme des deux poids.'[1]

§ 3. LAGRANGE's argument contains implicitly some *hypotheses of a statical nature,* regarding symmetry, addition of constraints,[2] etc.; and, in addition, it involves a geometrical property of the Euclidean triangle. But if we wish to omit the latter, a course which for certain reasons seems natural, the preceding conclusions will be modified.

Indeed, though we may still assume that the triangle ABD is in equilibrium about the point E, where the lines MN and CD intersect, we cannot assert that E is the middle point of CD, as this would be equivalent to assuming EUCLID's Postulate. Consequently, *we cannot assert that the single weight* 2 *P, applied at C, can be substituted for the two weights at A and B*, since, if such a change could take place, a lever would be in equilibrium, with equal weights at its ends, about a point which *cannot* be its middle point.

Vice versa, if we assume, with ARCHIMEDES, that two equal weights at the end can be replaced by a double weight at the middle point of the lever, then we can easily deduce that E is the middle point of CD, and from this it will follow that ABD is a Euclidean triangle.

Hence we have established the equivalence of Euclid's Fifth Postulate and the said hypothesis of Archimedes. Such equivalence is, of course, *relative* to the system of hypotheses which comprises, on the one hand, the above-named statical hypotheses, and, on the other, the ordinary geometrical hypotheses.

---

[1] *Oeuvres de Lagrange,* T. XI, p. 4—5.

[2] For an analysis of the *physical principles* on which ordinary statics is founded, cf. F. ENRIQUES, *Problemi della Scienza.* Cap. V. (Bologna, 1906). German translation, (Leipzig, 1910).

With the modern notation, we can speak of *forces*, *of the composition of forces*, of *resultants*, instead of *weights*, *levers*, etc.

Then the hypothesis referred to takes the following form:

The resultant of two equal forces in the same plane, applied at right angles to the extremities of a straight line and towards the same side of it, is a single force at the middle point of the line, of double the intensity of the given forces.

From what we have said above, if this law for the composition of forces were true, it would follow that the ordinary theory of parallels holds in space.

On the Composition of Forces Acting at a Point.

§ 4. The other fundamental principle of statics, *the law of the Parallelogram of Forces*, from the usual geometrical interpretation which it receives, is closely connected with the Euclidean nature of space. However, if we examine the essential part of this principle, namely, the analytical expression for the resultant R of two equal forces P, acting at a point, it is easy to show that it exists *independently* of any hypothesis on parallels.

This can be made clear by deducing the formula

$$R = 2P \cos \alpha,$$

where 2α is the angle formed by the two concurrent forces from the following principles:

1) Two or more forces, acting at the same point, have a definite resultant.

2) The resultant of two equal and opposite forces is zero.

3) The resultant of two or more forces, acting at a point, along the same straight line, is a force through the same point, equal to the sum of the given forces, and along the same line.

4) The resultant of two equal forces, acting at the same point, is directed along the line bisecting the angle between the two forces.

5) The magnitude of the resultant is a continuous function of the magnitude of the components.

Let us see briefly how we establish our theorem. The value R of the resultant of two forces of equal magnitude P, enclosing the angle 2α, is a function of P and α only.

Thus we can write
$$R = 2f(P, \alpha).$$

A first application of the principles named above shows that R is proportional to P, and this result is independent of any hypothesis on parallels [cf. note 1, p. 195]. Thus the preceding equation can be written more simply as
$$R = 2Pf(\alpha).$$

We now proceed to find the form of $f(\alpha)$.

§ 5. Let us calculate $f(\alpha)$ for some particular value of the angle.

(I) Let $\alpha = 45°$.

At the point O at which act the two forces P_1, P_2, of equal magnitude P, let us imagine two equal and opposite forces applied, perpendicular to R and of magnitude $\dfrac{R}{2}$ (cf. Fig. 64).

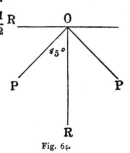
Fig. 64.

At the same time let us imagine R decomposed into two others, directed along R and of magnitude $\dfrac{R}{2}$.

We can then regard each force P as the resultant of two forces at right angles, of magnitude $\dfrac{R}{2}$.

We thus have

$$P = 2 \cdot \frac{R}{2} \cdot f(45°).$$

On the other hand, R being the resultant of P_1 and P_2, we have

$$R = 2\ Pf(45°).$$

From these two equations we obtain

$$f(45°) = \frac{1}{2}\ V\ \overline{2}.$$

(II)   Again let $α = 60°$.

In this case apply at O a force R' equal and opposite to R (cf. Fig. 65). The system of the two forces P and of R' is in equilibrium.

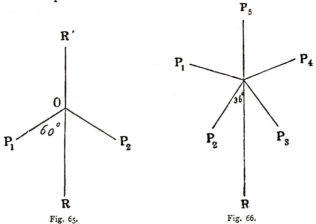

Fig. 65.   Fig. 66.

Thus by symmetry, $R' = P$.

Therefore, $R = P$.

But, on the other hand,

$$R = 2\ Pf(60°).$$

Therefore $f(60°) = \frac{1}{2}$.

(III)   Again let $α = 36°$.

At O let the five forces P_1, $P_2 .. P_5$, of magnitude P, be applied, such that each of them forms with the next an angle of $72°$ (cf. Fig. 66).

This system is in equilibrium.

For the resultant R of P_2 and P_3, we have

$$R = 2\,Pf\,(36°).$$

For the resultant R' of P_1 and P_4, we have

$$R' = 2\,Pf\,(72°).$$

On the other hand, R' has the same direction as P_5; that is, a direction opposite to that of R.

Therefore $2\,Pf\,(36°) = 2\,Pf\,(72°) + P.$

(1)   Therefore $2f\,(36°) = 2f\,(72°) + 1.$

If, instead, we take the resultants of P_1 and P_2, and of P_3 and P_4, we obtain two forces of magnitude $2\,Pf\,(36°)$, containing an angle of $144°$.

Taking the resultant of these two, we obtain a new force R'' of magnitude

$$4\,Pf\,(36°)\,f\,(72°).$$

Now R'', by the symmetry of the figure, has the same line of action as P_5, but acts in the opposite direction.

Thus, since equilibrium must exist,

$$P = 4\,Pf\,(36°)\,f\,(72°).$$

(2)   Therefore $1 = 4f\,(36°)\,f\,(72°).$

From the two equations (1) and (2) we obtain

$$f\,(36°) = \frac{1+\sqrt{5}}{4},\ f\,(72°) = \frac{-1+\sqrt{5}}{4},$$

on solving for $f\,(36°)$ and $f\,(72°)$.

§ 6. By arguments similar to those used in the preceding section we could deduce other values for $f\,(\alpha)$. However, if we restrict ourselves only to those just found, and compare them with the corresponding values of cos α, we obtain the following table:

$$\cos 0° = 1 \qquad f(0°) = 1$$

$$\cos 36° = \frac{1 + \sqrt{5}}{4} \qquad f(36°) = \frac{1 + \sqrt{5}}{4}$$

$$\cos 45° = \frac{\sqrt{2}}{2} \qquad f(45°) = \frac{\sqrt{2}}{2}$$

$$\cos 60° = \frac{1}{2} \qquad f(60°) = \frac{1}{2}$$

$$\cos 72° = \frac{-1 + \sqrt{5}}{4} \qquad f(72°) = \frac{-1 + \sqrt{5}}{4}$$

$$\cos 90° = 0 \qquad f(90°) = 0.$$

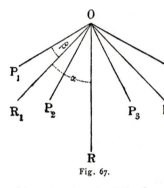

Fig. 67.

This table suggests the identity of the two functions $f(\alpha)$ and cos α. For fuller confirmation of this fact, we determine the functional equation which $f(\alpha)$ satisfies (cf. Fig. 67).

To this end let us consider four forces P_1, P_2, P_3, P_4 of magnitude P, acting at one point, forming with each other the following angles

$$\measuredangle P_1 P_2 = \measuredangle P_3 P_4 = 2\beta$$

$$\measuredangle P_2 P_3 = 2(\alpha - \beta)$$

$$\measuredangle P_1 P_4 = 2(\alpha + \beta).$$

We shall determine the resultant R of these four forces in two different ways.

Taking P_1 with P_2, and P_3 with P_4 we obtain two forces R_1 and R_2, of magnitude

$$2 P f(\beta),$$

inclined at an angle 2β. Taking the resultant of R_1 and R_2, we have a force R, such that

$$R = 4\,Pf(\alpha)\,f(\beta).$$

On the other hand, taking P_1 with P_4, and P_2 with P_3, we obtain two resultants, both along the direction of R, and of magnitudes

$$2\,Pf(\alpha + \beta),\ 2\,Pf(\alpha - \beta),$$

respectively.

These two forces have a resultant equal to their sum, and thus

$$R = 2\,P f(\alpha + \beta) + 2\,P f(\alpha - \beta).$$

Comparing the two values of R, we find that

(1) $$2f(\alpha)\,f(\beta) = f(\alpha + \beta) + f(\alpha - \beta)$$

is the functional equation required.

If we now remember that

$$\cos(\alpha + \beta) + \cos(\alpha - \beta) = 2\cos\alpha\cos\beta,$$

and take account of the identity between $f(\alpha)$ and $\cos\alpha$ in the preceding table for certain values of α, and the hypothesis that $f(\alpha)$ is continuous, without further argument we can write

$$f(\alpha) = \cos\alpha.$$

It follows that

$$R = 2\,P\cos a.$$

The validity of this formula of the Euclidean space is thus also established for the Non-Euclidean spaces.

§ 7. The law of composition of two equal concurrent forces leads to the solution of the general problem of the resultant, since we can assign, without any further hypothesis, the components of a force R along two rectangular axes through its point of application O.

Let the two perpendicular lines be taken as the axes of x and y, and let R make the angles α, β with them (cf. Fig. 68).

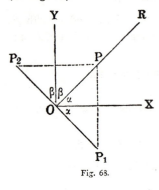

Fig. 68.

Through O draw the line which makes an angle α with Ox and an angle β with Oy. Imagine two equal and opposite forces P_1 and P_2 to act along this line at O, their magnitude being $\frac{R}{2}$. Also imagine the force R replaced by the two equal forces P, of magnitude $\frac{R}{2}$, acting in the same direction as R.

Then the system P_1, P_2, P, P has R for resultant. But P_1 and P, taken together, have a resultant

$$X = R \cos \alpha$$

along Ox: and P_2 and P, taken together, have a resultant

$$Y = R \cos \beta$$

along Oy.

These two forces are the components of R along the two perpendicular lines. As to their magnitudes, they are identical with what we would obtain in the ordinary theory founded upon the principle of the Parallelogram of Forces. However, the lines OX and OY, which represent the components upon the axes, *are not necessarily the projections of R*, as in the Euclidean case. Indeed we can easily see that, if these lines were the orthogonal projections of R upon the axes, the Euclidean Hypothesis would hold in the plane.

§ 8. The functional method applied in § 6 to the composition of two equal forces acting at a point, is derived from D. DE FONCENEX [1734—1799]. By a method analogous to that which led us to the equation for $f(\alpha)$ ($= y$), FONCENEX arrived at the differential equation [1]

$$\frac{d^2y}{d\alpha^2} + k^2y = 0.$$

From this, on integrating and taking account of the initial conditions of the problem, he obtained the known expression for $f(\alpha)$.

However the application of the principles of the Infinitesimal Calculus, requires the continuity and differentiability of $f(\alpha)$, conditions, which, as FONCENEX remarks, involve the (physical) nature of the problem. But as he wishes to go 'jusqu'aux difficultés les moins fondées', he avails himself of the Calculus of Finite Differences, and of a Difference Equation, which allows him to obtain $f(\alpha)$ for all values of α which are commensurable with π. The case α incommensurable is treated 'par une méthode familière aux Géomètres et frequente surtout le écrits des Anciens'; that is, by the Method of Exhaustion. [2]

All FONCENEX' argument, and therefore that given in

---

[1] We could obtain this equation from (1) p. 189 as follows: Put $\beta = d\alpha$ and suppose that $f(\alpha)$ can be expanded by TAYLOR's Series for every value of α.

Then we have

$$2f(\alpha)\left(f(0) + d\alpha\, f'(0) + \frac{d\alpha^2}{2} f''(0) \ldots\right)$$
$$= 2f(\alpha) + 2\frac{d\alpha^2}{2}f''(\alpha) + \ldots$$

Equating the coefficients of $d\alpha^2$ and putting $y = f(\alpha)$ and $k^2 = -f''(0)$, we have

$$\frac{d^2y}{d\alpha^2} + k^2y = 0.$$

[2] Cf. FONCENEX: *Sur les principes fondamentaux de la Mécanique.* Misc. Taurinensia. T. II, p. 305—315 (1760—1761). His argument is repeated and explained by A. GENOCCHI in his paper: *Sur un Mémoire de Daviet de Foncenex et sur les géométries non-euclidiennes.* Torino, Memorie (2), T. XXIX, p. 366—371 (1877).

§ 6, is independent of EUCLID's Postulate. However, it should be remarked that FONCENEX' aim was not to make the law of composition of concurrent forces independent of the theory of parallels, but rather to *prove* the law itself. Probably he held, as other geometers [D. BERNOUILLI, D'ALEMBERT], that it was a truth independent of any experimental foundation.

Non-Euclidean Statics.

§ 9. Having thus shown that the analytical law for the composition of concurrent forces does not depend on EUCLID's Fifth Postulate, we proceed to deduce the law according to which forces perpendicular to a line will be composed.

Let A, A' be the points of application of two torces P_1, P_2 of equal magnitude P (cf. Fig. 69).

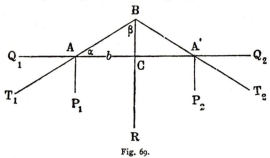

Fig. 69.

Let C be the middle point of AA', and B a point on the perpendicular BC to AA'.

Joining AB and $A'B$, and putting

$$\sphericalangle BAC = \alpha, \quad \sphericalangle ABC = \beta,$$

it is clear that the force P_1 can be regarded as a component of a force T_1, acting at A and along BA.

The magnitude of this force is given by

$$T = \frac{P}{\sin \alpha}.$$

The other component Q_1, at right angles to P_1, is given by

$$Q = T \cos \alpha = P \cot \alpha.$$

Repeating this process with the force P_2, we obtain the following system of coplanar forces:

(1) System P_1, P_2.

(2) System P_1, P_2, Q_1, Q_2.

(3) System T_1, T_2.

If we *assume that* we can move the point of application of a force along its line of action, it is clear that the first two systems are equivalent, and because (2) is equivalent to (3), we can substitute for the two forces P_1, P_2, the two forces T_1 and T_2.

The latter, being moved along their lines of action to B, can be composed into one force

$$R = 2\,T \cos \beta = 2\,P\,\frac{\cos \beta}{\sin \alpha}.$$

This, in its turn, can be moved to C, its direction perpendicular to AA' remaining unchanged.

This result, which is obviously *independent* of EUCLID's Postulate, can be applied to the three systems of geometry:

Euclid's Geometry.

In the triangle ABC we have

$$\cos \beta = \sin \alpha.$$

Therefore

$$R = 2\,P.$$

Geometry of Lobatschewsky-Bolyai.

In the triangle ABC, if we denote the side AA' by $2\,b$, we have

$$\frac{\cos \beta}{\sin \alpha} = \cosh \frac{b}{k} \quad (\text{p. } 117).$$

Thus

$$R = 2\,P \cosh \frac{b}{k}.$$

Riemann's Geometry.

In the same triangle we have

$$\frac{\cos \beta}{\sin \alpha} = \cos \frac{b}{k}.$$

Therefore

$$R = 2\,P \cos \frac{b}{k}.$$

Conclusion.

It is only in Euclidean space that the resultant of two equal forces, perpendicular to the same line, is equal to the sum of the two given forces. In the Non-Euclidean spaces the resultant depends, in the manner indicated above, on the distance between the points at which the two forces are applied.[1]

§ 10. The case of two unequal forces P, Q, perpendicular to the same straight line, is treated in a similar manner.

In the Euclidean Geometry we obtain the known results;

$$R = P + Q,$$
$$\frac{R}{p+q} = \frac{P}{q} = \frac{Q}{p}.$$

In the Geometry of LOBATSCHEWSKY-BOLYAI the problem of the resultant leads to the following equations:

$$R = P \cosh \frac{p}{k} + Q \cosh \frac{q}{k},$$
$$\frac{R}{\sinh \frac{p+q}{k}} = \frac{P}{\sinh \frac{q}{k}} = \frac{Q}{\sinh \frac{p}{k}}.$$

Then, by the usual substitution of the circular functions for the hyperbolic, we obtain the corresponding result for RIEMANN's Geometry:

---

[1] For a fuller treatment of Non-Euclidean Statics, the reader is referred to the following authors: J. M. DE TILLY, *Études de Mécanique abstraite,* Mém. couronnés et autres mém., T. XXI (1870). J. ANDRADE, *La Statique et les Géométries de Lobatschewsky, d'Euclide, et de Riemann.* Appendix (II) of the work quoted on p. 181.

$$R = P \cos \frac{p}{k} + Q \cos \frac{q}{k},$$

$$\frac{R}{\sin \dfrac{p+q}{k}} = \frac{P}{\sin \dfrac{q}{k}} = \frac{Q}{\sin \dfrac{p}{k}}.$$

In these formulæ p, q, denote the distances of the points of application of P and Q from that of R.

These results can be summed up in a single formula, valid for *Absolute Geometry*;

$$R = P.\, Ep + Q.\, Eq,$$

$$\frac{R}{\odot (p+q)} = \frac{P}{\bigcirc(q)} = \frac{Q}{\bigcirc(p)}.$$

To obtain these results directly, it is sufficient to use the formulæ of Absolute Trigonometry, instead of the Euclidean or Non-Euclidean, in the argument of which a sketch has just been given.

Deduction of Plane Trigonometry from Statics.

§ 11. Let us see, in conclusion, how it is possible to treat the converse question: *given the law of composition of forces, to deduce the fundamental equations of trigonometry.*

To this end we note that the magnitude of the resultant R of two equal forces P, perpendicular to a line AA' of length $2\,b$, will in general be a function of P and b.

Denoting this function by

$$\varphi\, (P,\, b),$$

we have

$$R = \varphi\, (P,\, b),$$

or more simply[1]

$$R = P\varphi\, (b).$$

---

[1] The proportionality of R and P follows from the *law of association* on which the composition of forces depends. In fact, let us imagine each of the forces P, acting at A and A', to be

On the other hand in § 9 (p. 193), we were brought to the following expression for R:

$$R = 2\,P\frac{\cos\beta}{\sin\alpha}.$$

Eliminating R and P, between these, we have

$$\varphi\,(b) = \frac{\cos\beta}{\sin\alpha}.$$

Thus if the analytical expression for $\varphi\,(b)$ is known, this formula will supply a relation between the sides and angles of a right-angled triangle.

To determine $\varphi\,(b)$, it is necessary to establish the corresponding functional equation.

With this view, let us apply perpendicularly to the line AA', the four equal forces $P_1,\,P_2,\,P_3,\,P_4$, in such a way that the points of application of P_1 and P_4, P_2 and P_3, are distant $2\,(a+b)$ and $2\,(b-a)$, respectively (cf. Fig. 70).

We can determine the resultant R of these four forces in two different ways:

(1) Taking P_1 with P_2, and P_3 with P_4, we obtain two forces $R_1,\,R_2$ of magnitude:

$$P\varphi\,(a);$$

---

replaced by n equal forces, applied at A and A'. Combining these, we would have for R the expression

$$R = n\,\varphi\left(\frac{P}{n},\,b\right).$$

Comparing this result with the equation given in the text, we have

$$\varphi\left(\frac{P}{n},\,b\right) = \frac{1}{n}\,\varphi\,(P,\,b).$$

Similarly we have

$$\varphi\,(kP,\,b) = k\varphi\,(P,\,b),$$

for every rational value of k; and the formula may be extended to irrational values.

Then putting $P=1$ and $k=P$ we obtain

$$\varphi\,(P,\,b) = P\,\varphi\,(b). \qquad \text{Q. E. D.}$$

and taking R_1, R_2 together, we obtain
$$R = P\varphi(a)\varphi(b).$$

(ii) Taking P_1 with P_4, we obtain a force of magnitude:
$$P\varphi(b+a),$$
and taking P_2 with P_3, we obtain another of magnitude:
$$P\varphi(b-a).$$

Taking these two together we have, finally,
$$R = P\varphi(b+a) + P\varphi(b-a).$$

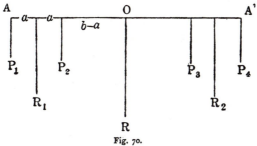

Fig. 70.

From the two expressions for R we obtain the functional equation which $\varphi(b)$ satisfies, namely,

(2) $$\varphi(b)\varphi(a) = \varphi(b+a) + \varphi(b-a).$$

This equation, if we put $\varphi(b) = 2f(b)$, is identical with that met in § 6 (p. 189), in treating the composition of concurrent forces.

The method followed in finding (2) is due to D'ALEMBERT.[1] However, if we suppose a and b equal to each other, and if we note that $\varphi(o) = 2$, the equation reduces to

(3) $$[\varphi(x)]^2 = \varphi(2x) + 2.$$

This last equation was obtained previously by FONCENEX, in connection with the equilibrium of the lever.[2]

---

[1] *Opuscules mathématiques*, T. VI, p. 371 (1779).

[2] Cf. p. 319—322 of the work by FONCENEX, referred to above.

§ 12.   The statical problem of the composition of forces is thus reduced to the integration of a functional equation.

FONCENEX, who was the first to treat it in this way[1], thought that the only solution of (3), was $\varphi(x) = $ const. If this were so, the constant would be 2, as is easily verified.

Later LAPLACE and D'ALEMBERT integrated (3), obtaining

$$\varphi(x) = e^{\frac{x}{c}} + e^{-\frac{x}{c}},$$

where c is a constant, or any function which takes the same value when x is changed to $2\,x$.[2]

The solution of LAPLACE and D'ALEMBERT, applied to the statical problem of the preceding section, leads to the case in which c is a function of x. Further, since we cannot admit values of c such as $a + i\,b$, where a, b are both different from zero, we have three possible cases, according as c is real, a pure imaginary, or infinite.[3] Corresponding to these

---

[1] We have stated above (p. 53), when speaking of FONCENEX' memoir, that, if it was not the work of Lagrange, it was certainly inspired by him.   This opinion, accepted by GENOCCHI and other geometers, dates from DELAMBRE.   The distinguished biographer of LAGRANGE puts the matter in the following words: "*Il (Lagrange) fournissait à Foncenex la partie analytique de ses mémoires en lui laissant le soin de développer les raisonnements sur lesquels portaient ses formules.   En effet, on remarque déjà dans ces mémoires* (of FONCENEX) *cette marche purement analitique, qui depuis a fait le caractère des grandes productions de Lagrange.   Il avait trouvé une nouvelle théorie du levier*". *Notices sur la voie et les ouvrages de M. le Comte Lagrange.* Mém. Inst. de France, classe Math. et Physique, T. XIII, p. XXXV (1812).

[2] Cf. D'ALEMBERT: *Sur les principes de la Mécanique:* Mém. de l'Ac. des Sciences de Paris (1769). — LAPLACE: *Recherches sur l'intégration des équations différentielles:* Mém. Ac. sciences de Paris (savants étrangers) T. VII (1733). *Oeuvres de Laplace,* T. VIII, p. 106—7.

[3] We can obtain this result directly by integrating the equa-

three cases, we have three possible laws for the composition of forces, and consequently three distinct types of equations connecting the sides and angles of a triangle. These results are brought together in the following table, where k denotes a real positive number.

Value of c	Form of $\varphi(x)$	Trigonometrical equations	Nature of plane
$c = k$	$e^{\frac{x}{k}} + e^{-\frac{x}{k}} = 2\cosh\frac{x}{k}$	$\cosh\frac{b}{k} = \frac{\cos\beta}{\sin\alpha}$	hyperbolic
$c = ik$	$e^{\frac{ix}{k}} + e^{-\frac{ix}{k}} = 2\cos\frac{x}{k}$	$\cos\frac{b}{k} = \frac{\cos\beta}{\sin\alpha}$	elliptic
$c = \infty$	$e^{\frac{x}{\infty}} + e^{-\frac{x}{\infty}} = 2$	$1 = \frac{\cos\beta}{\sin\alpha}$	parabolic

Conclusion: The law for the composition of forces perpendicular to a straight line, leads, in a certain sense, to the relations which hold between the sides and angles of a triangle, and thus to the geometrical properties of the plane and of space.

This fact was completely established by A. GENOCCHI [1817—1889] in two most important papers[1], to which the reader is referred for full historical and bibliographical notes upon this question.

---

tion (2), or, what amounts to the same thing, equation (1) of § 6. Cf., for this, the elementary method employed by CAUCHY for finding the function satisfying (1). *Oeuvres de Cauchy*, (sér. 2). T. III, p. 106—113.

[1] One of them is the Memoir referred to on p. 191. The other, which dates from 1869, is entitled: *Dei primi principii della meccanica e della geometria in relazione al postulato d'Euclide*. Annali della Società italiana delle Scienze (3). T. II, p. 153—189.

Appendix II.

Clifford's Parallels and Surface.
Sketch of Clifford-Klein's Problem.

Clifford's Parallels.

§ 1. EUCLID's Parallels are straight lines possessing the following properties:

a) They are coplanar.

b) They have no common points.

c) They are equidistant.

If we give up the condition (c) and adopt the views of GAUSS, LOBATSCHEWSKY and BOLYAI, we obtain a first extension of the notion of parallelism. But the parallels which correspond to it have very few properties in common with the ordinary parallels. This is due to the fact that the most beautiful properties we meet in studying the latter depend principally on the condition (c). For this reason we are led to seek such an extension of the notion of parallelism, that, so far as possible, the *new parallels* shall still possess the characteristics, which, in Euclidean geometry, depend on their equidistance. Thus, following W. K. CLIFFORD [1845—1879], we give up the *property of coplanarity*, in the definition of parallels, and retain the other two. The new definition of parallels will be as follows:

Two straight lines, in the same or in different planes, are called parallel, when the points of the one are equidistant from the points of the other.

§ **2.** Two cases, then, present themselves, according as these parallels lie, or do not lie, in the same plane.

The case in which the equidistant straight lines are coplanar is quickly exhausted, since the discussion in the earlier part of this book [§ 8] allows us to state that the corresponding space is the ordinary Euclidean. We shall, therefore, suppose that the two equidistant straight lines r and s are not in the same plane, and that the perpendiculars drawn from r to s are equal. Obviously these lines will also be perpendicular to r. Let AA', BB'

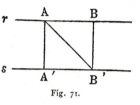

Fig. 71.

be two such perpendiculars (Fig. 71). The skew quadrilateral $ABB'A'$, which is thus obtained, has its four angles and two opposite sides equal. It is easy to see that the other two opposite sides AB, $A'B'$ are equal, and that the interior alternate angles, which each diagonal—e. g. AB'—makes with the two parallels, are equal. This follows from the congruence of the two right-angled triangles $AA'B'$ and ABB'.

If now we examine the solid angle at A, from a theorem valid in all the three geometrical systems, we can write

$$\angle A'AB' + \angle B'AB > \angle A'AB = 1 \text{ right angle.}$$

This inequality, taken along with the fact that the angles $AB'A'$ and $B'AB$ are equal, can be written thus:

$$\angle A'AB' + \angle AB'A' > 1 \text{ right angle.}$$

Stated in this way, we see that the sum of the acute angles in the right-angled triangle $AA'B'$ is greater than a right angle. Thus in the said triangle the *Hypothesis of the Obtuse Angle* is verified, and consequently *parallels not in the same plane can exist only in the space of Riemann.*

§ 3. Now to prove that in the *elliptic* space of RIEMANN
there actually do exist pairs of straight lines, not in the same
plane and equidistant, let us consider an arbitrary straight
line *r* and the infinite number of planes perpendicular to it.
These planes all pass through another line *r'*, the polar
of *r* in the absolute polarity of the elliptic space. Any line
whatever, joining a point of *r* with a point of *r'*, is perpendicular both to *r* and to *r'*, and has a constant length, equal
to half the length of a straight line. From this it follows
that *r*, *r'* are two equidistant straight lines, not in the same
plane.

But two such equidistants represent a very particular
case, since all the points of *r* have the same distance not
only from *r'*, but from *all the points of r'*.

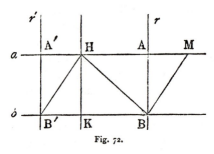

Fig. 72.

To establish the existence of straight lines in which the
last peculiarity does not exist, we consider again two lines
r and *r'*, one of which is the polar of the other (Fig. 72).
Upon these let the equal segments *AB*, *A'B'* be taken, each
less than half the length of a straight line. Joining *A* with
A', and *B* with *B'*, we obtain two straight lines *a*, *b*, not
polar the one to the other, and both perpendicular to the
lines *r*, *r'*.

It can easily be proved that *a*, *b* are equidistant. To
show this, take a segment *A'H* upon *AA'*; then on the

supplementary line [1] to $A'HA$, take the segment AM equal to $A'H$. If the points H and M are joined respectively with B' and B, we obtain two right-angled triangles $A'B'H$, ABM, which, in consequence of our construction, are congruent.

We thus have the equality

$$HB' = BM.$$

Now if H and B are joined, and the two triangles HBB' and HBM are compared, we see immediately that they are equal. They have the side HB common, the sides HB' and MB equal, by the preceding result, and finally BB' and HM are also equal, each being half of a straight line.

This means, in other words, that the various points of the straight line a are equidistant from the line b. Now since the argument can be repeated, starting from the line b and dropping the perpendiculars to a, we conclude that the line HK, in addition to being perpendicular to b, is also perpendicular to a.

We remark, further, that from the equality of the various segments AB, HK, $A'B'$, ... the equality of the respective supplementary segments is deduced, so that the two lines a, b, can be regarded as equidistant the one from the other, in two different ways. If then it happened that the line AB were equal to its supplement, we would have the ex ceptional case, which we noted previously, where a, b are the polars of each other, and consequently all the points of a are equidistant from the different points of b.

§ 4. The non-planar parallels of elliptic space were discovered by CLIFFORD in 1873.[2] Their most remarkable properties are as follows:

---

[1] The two different segments, determined by two points on a straight line, are called supplementary.

[2] *Preliminary Sketch of Biquaternions*. Proc. Lond. Math. Soc. Vol. IV. p. 381—395 (1873). *Clifford's Mathematical Papers*, p. 181—200.

(i) *If a straight line meets two parallels, it makes with them equal corresponding angles, equal interior alternate angles, etc.*

(ii) *If in a skew quadrilateral the opposite sides are equal and the adjacent angles supplementary, then the opposite sides are parallel.*

Such a quadrilateral can therefore be called a *skew parallelogram.*

The first of these two theorems can be immediately verified; the second can be proved by a similar argument to that employed in § 3.

(iii) *If two straight lines are equal and parallel, and their extremities are suitably joined, we obtain a skew parallelogram.*

This result, which can be looked upon, in a certain sense, as the converse of (ii), can also be readily established.

(iv) *Through any point (M) in space, which does not lie on the polar of a straight line (r), two parallels can be drawn to that line.*

Indeed, let the perpendicular *MN* be drawn from *M* to *r*, and let *N'* be the point in which the polar of *MN*

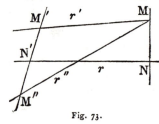

Fig. 73.

meets *r* (Fig. 73). From this polar cut off the two segments *N'M'*, *N'M''*, equal to *NM*, and join the points *M'*, *M''* to *M*. The two lines *r'*, *r''*, thus obtained, are the required parallels.

If *M* lay on the polar of *r*, then *MN* would be equal to half the straight line; the two points *M'*, *M''* would coincide: and the two parallels *r'*, *r''* would also coincide.

The angle between the two parallels r', r'' can be measured by the segment $M'M''$, which the two arms of the angle intercept on the polar of its vertex. In this way we can say that half of the angle between r' and r'', that is, the *angle of parallelism*, is equal to the *distance of parallelism*.

To distinguish the two parallels r', r'', let us consider a helicoidal movement of space, with MN for axis, in which the pencil of planes perpendicular to MN, and the axis $M'M''$ of that pencil, obviously remain fixed. Such a movement can be considered as the resultant of a translation along MN, accompanied by a rotation about the same axis: or by two translations, one along MN, the other along $M'M''$. If the two translations are of equal amount, we obtain a space *vector*.

Vectors can be *right-handed* or *left-handed*. Thus, referring to the two parallels r', r'', it is clear that one of them will be superposed upon r by a right-handed vector of magnitude MN, while the other will be superposed on r by a left-handed vector of the same magnitude. Of the two lines r', r'', one could be called the *right-handed parallel* and the other the *left-handed parallel* to r.

(v) *Two right-handed (or left-handed) parallels to a straight line are right-handed (or left-handed) parallels to each other.*

Let b, c be two right-handed parallels to a. From the two points A, A' of a, distant from each other half the length of a straight line, draw the perpendiculars AB, $A'B'$ on b, and the perpendiculars AC, $A'C'$ on c (cf. Fig. 74).

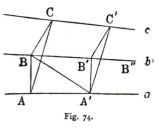

Fig. 74.

The lines $A'B'$, $A'C'$ are the polars of AB and AC. Therefore $\angle BAC = \angle B'A'C'$.

Further, by the properties of parallels
$$AB = A'B', \qquad AC = A'C'.$$
Therefore the triangles ABC, $A'B'C'$ are equal
Thus it follows that
$$BC = B'C'.$$
Again, since
$$BB' = AA' = CC',$$
the skew quadrilateral $BB'C'C$ has its opposite sides equal.

But to establish the parallelism of b, c, we must also prove that the adjacent angles of the said quadrilateral are supplementary (cf. ii). For this we compare the two solid angles B $(AB'C)$ and B' $(AB''C')$. In these the following relations hold:
$$\angle ABB' = \angle A'B'B'' = \text{1 right angle}$$
$$\angle ABC = \angle A'B'C'.$$

Further, the two dihedral angles, which have BA and $B'A'$ for their edges, are each equal to a right angle, diminished (or increased) by the dihedral angle whose normal section is the angle $A'BB'$.

Therefore the said two solid angles are equal. From this the equality of the two angles $B'BC$, $B''B'C'$ follows. Hence we can prove that the angles B, B' of the quadrilateral $BB'C'C$ are supplementary, and then (on drawing the diagonals of the quadrilateral, etc.) that the angle B is supplementary to C, and C supplementary to C', etc.

Thus b and c are parallel. From the figure it is clear that the parallelism between b and c is right-handed, if that is the nature of the parallelism between the said lines and the line a.

Clifford's Surface.

§ 5. From the preceding argument it follows *that all the lines which meet three right-handed parallels are left-handed parallels to each other.*

Indeed, if ABC is a transversal cutting the three lines a, b, c, and if three equal segments AA', BB', CC' are taken on these lines in the same direction,[1] the points $A'B'C'$ lie on a line parallel to ABC. The parallelism between ABC and $A'B'C'$ is thus left-handed.

From this we deduce that three parallels a, b, c, define a ruled surface of the second order (CLIFFORD's Surface). On this surface the lines cutting a, b, c form one system of generators (g_s): the second system of generators (g_d) is formed by the infinite number of lines, which, like a, b, c, meet (g_s).

CLIFFORD's Surface possesses the following characteristic properties:

a) *Two generators of the same system are parallel to each other.*

b) *Two generators of opposite systems cut each other at a constant angle.*

§ 6. We proceed to show that *Clifford's Surface has two distinct axes of revolution.*

To prove this, from any point M draw the parallels d (right-handed), s (left-handed), to a line r, and denote by δ the distance MN of each parallel from r (cf. Fig. 75).

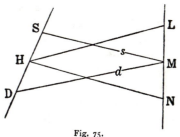

Fig. 75.

Keeping d fixed, let s rotate about r, and let s', s'', s''' ... be the successive positions which s takes in this rotation.

---

[1] It is clear that if a direction is fixed for one line, it is then fixed for every line parallel to the first.

It is clear that s, s', s'' . . . are all left-handed parallels to r and that all intersect the line d.

Thus s in its rotation about r generates a CLIFFORD'S Surface.

Vice versa, if d and s are two generators of a CLIFFORD'S Surface, which pass through a point M of the surface, and 2 δ the angle between them, we can raise the perpendicular to the plane sd at M and upon it cut off the lines $ML = MN = δ$.

Let D and S be the points where the polar of LN meets the lines d and s, respectively, and let H be the middle point of $DS = 2$ δ.

Then the lines HL and HN are parallel, both to s and d.

Of the two lines HL and HN choose that which is a right-handed parallel to d and a left-handed parallel to s, say the line HN.

Then the given CLIFFORD'S Surface can be generated by the revolution of s or d about HN.

In this way it is proved that every CLIFFORD'S Surface possesses *one axis of rotation* and that every point on the surface is equidistant from it.

The existence of another axis of rotation follows immediately, if we remember that all the points of space, equidistant from HN, are also equidistant from the line which is the polar of HN.

This line will, therefore, be the *second axis of rotation* of the CLIFFORD'S Surface.

§ 7. The equidistance of the points of CLIFFORD'S Surface from each axis of rotation leads to another most remarkable property of the surfaces. In fact, every plane passing through an axis r intersects it in a line equidistant from the axis. The points of this line, being also equally distant from the point (O) in which the plane of section meets the other axis of the surface, lie on a circle, whose centre (O) is the pole of r with respect to the said line. Therefore the *meridians* and the *parallels* of the surface are circles.

The surface can thus be generated by making a circle rotate about the polar of its centre, or by making a circle move so that its centre describes a straight line, while its plane is maintained constantly perpendicular to it (BIANCHI).[1]

This last method of generating the surface, common also to the Euclidean cylinder, brings out the analogy between CLIFFORD's Surface and the ordinary circular cylinder This analogy could be carried further, by considering the properties of the helicoidal paths of the points of the surface, when the space is submitted to a screwing motion about either of the axes of the surface.

§ **8.** Finally, we shall show that the geometry on CLIFFORD's Surface, understood in the sense explained in §§ 67, 68, is identical with Euclidean geometry.

To prove this, let us determine the law according to which the element of distance between two points on the surface is measured.

Let u, v, be respectively a parallel and a meridian through a point O on the surface, and M any arbitrary point upon it.

Let the meridian and parallel through M cut off the arcs OP, OQ from u and v. The lengths u, v of these arcs will be the coordinates of M. The analogy between the system of coordinates here adopted and the Cartesian orthogonal system is evident (cf. Fig. 76).

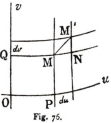

Fig. 76.

[1] *Sulla superficie a curvatura nulla in geometria ellittica.* Ann. di Mat. (2) XXIV, p. 107 (1896). Also *Lezioni di Geometria Differenziale.* 2a Ed., Vol. I, p. 454 (Pisa, 1902).

Let M' be a point whose distance from M is infinitesimal. If (u, v) are the coordinates of M, we can take $(u + du, v + dv)$ for those of M'.

Now consider the infinitesimal triangle $MM'N$, whose third vertex N is the point in which the parallel through M intersects the meridian through M'. It is clear that the angle MNM' is a right angle, and that the sides MN, NM' are equal to du, dv.

On the other hand, this triangle can be regarded as rectilinear (as it lies on the tangent plane at M). So that, from the properties of infinitesimal plane triangles, its hypotenuse and its sides, by the Theorem of Pythagoras, are connected by the relation

$$ds^2 = du^2 + dv^2.$$

But this expression for ds^2 is characteristic of ordinary geometry, so that we can immediately deduce *that the properties of the Euclidean plane hold in every normal region on a Clifford's Surface.*

An important application of this result leads to the evaluation of the *area* of this surface. Indeed, if we break it up into such congruent infinitesimal parallelograms by means of its generators, the area of one of these will be given by the ordinary expression

$$dx \, dy \sin \theta,$$

where dx, dy are the lengths of the sides and θ is the constant angle between them (the angle between two generators).

The area of the surface is therefore

$$\Sigma \, dx \, dy \sin \theta = \sin \theta \, \Sigma \, dx \cdot \Sigma \, dy.$$

But both the sums $\Sigma \, dx$, $\Sigma \, dy$ represent the length l of a straight line.

Therefore the area Δ of Clifford's Surface takes the very simple form,

$$\Delta = l^2 \sin \theta,$$

which is identical with the expression for the area of a Euclidean parallelogram (CLIFFORD). [1]

Sketch of Clifford-Klein's Problem.

§ 9. CLIFFORD'S ideas, explained in the preceding sections, led KLEIN to a new statement of the fundamental problem of geometry.

In giving a short sketch of KLEIN's views, let us refer to the results of § 68 regarding the possibility of interpreting plane geometry by that on the surfaces of constant curvature. The contrast between the properties of the Euclidean and Non-Euclidean planes and those of the said surfaces was there restricted to suitably bounded regions. In extending the comparison to the *unbounded regions*, we are met, in general, by differences; in some cases due to the presence of *singular points* on the surfaces (e. g., vertex of a cone); in others, to the different *connectivities* of the surfaces.

Leaving aside the singular points, let us take the circular cylinder as an example of a surface of constant curvature, *everywhere regular*, but possessed of a connectivity different from that of the Euclidean plane.

The difference between the geometry of the plane and that of the cylinder, both understood in the complete sense, has been already noticed on p. 140, where it was observed that the postulate of congruence between two arbitrary straight lines ceases to be true on the cylinder. Nevertheless there are numerous properties common to the two geometries,

---

[1] *Preliminary Sketch*, cf. p. 203 above. The properties of this surface were referred to only very briefly by CLIFFORD in 1873. They are developed more fully by KLEIN in his memoir: *Zur nicht-euklidischen Geometrie*, Math. Ann. Bd. XXXVII, p. 544—572 (1890).

which have their origin in the double characteristic, that both the plane and the cylinder have the same curvature, and that they are both regular.

These properties can be summarized thus:

1) The geometry of *any* normal region of the cylinder is identical with that of *any* normal region of the plane.

2) The geometry of any normal region *whatsoever* of the cylinder, fixed with respect to an arbitrary point upon it, is identical with the geometry of any normal region *whatsoever* of the plane.

The importance of the comparison between the geometry of the plane and that of a surface, founded on the properties (1) and (2), arises from the following considerations:

A geometry of the plane, based upon experimental criteria, depends on two distinct groups of hypotheses. The first group expresses the validity of certain facts, directly observed in a region accessible to experiment (*postulates of the normal region*); the second group extends to inaccessible regions some properties of the initial region (*postulates of extension*).

The postulates of extension could demand, e. g., that the properties of the accessible region should be valid in the entire plane. We would then be brought to the two forms, the parabolic and the hyperbolic plane. If, on the other hand, the said postulates demanded the extension of these properties, with the exception of that which attributes to the straight line the character of an open line, we ought to take account of the elliptic plane as well as the two planes mentioned.

But the preceding discussion on the regular surfaces of constant curvature suggests a more general method of enunciating the postulates of extension. We might, indeed, simply demand that the properties of the initial region should hold in the neighbourhood of every point of the plane. In this case, the class of possible forms of planes receives considerable additions. We could, e. g., conceive a form with zero curvature, of double connectivity, and able to be completely represented on the cylinder of Euclidean space.

The object of Clifford-Klein's problem is the determination of all the two dimensional manifolds of constant curvature, which are everywhere regular.

§ 10. Is it possible to realise, with suitable regular surfaces of constant curvature, in the Euclidean space, all the *forms* of CLIFFORD-KLEIN?

The answer is in the negative, as the following example clearly shows. The only regular *developable* surface of the Euclidean space, whose geometry is not identical with that of the plane, is the cylinder with closed cross-section. On the other hand, CLIFFORD's Surface in the elliptic space is a regular surface of zero curvature, which is essentially different from the plane and cylinder.

However with suitable *conventions* we can *represent* CLIFFORD's Surface even in ordinary space.

Let us return again to the cylinder. If we wish to *unfold* the cylinder, we must first render it *simply connected* by a cut along a generator (g); then, by bending without stretching, it can be spread out on the plane, covering a *strip* between two parallels (g_1, g_2).

There is a one-one correspondence between the points of the cylinder and those of the strip. The only exception is afforded by the points of the generator (g), to each of which correspond two points, situated the one on g_1, the other on g_2. However, if it is *agreed* to regard these two points as *identical*, that is, as a single point, then the correspondence becomes one-one without exception, and *the geometry of the strip is completely identical with that of the cylinder.*

A representation analogous to the above can also be adopted for CLIFFORD's Surface. First the surface is made simply connected by two cuts along the intersecting generators (g, g'). In this way a skew parallelogram is obtained in the elliptic space. Its sides have each the length of a straight line, and its angles θ and θ' [θ + θ' = 2 right angles] are the angles between g and g'.

This being done, we take a rhombus in the Euclidean plane, whose sides are the length of the straight line in the elliptic plane, and whose angles are θ, θ'. On this rhombus CLIFFORD's Surface can be represented *congruently* (developed). The correspondence between the points of the surface and those of the rhombus is a one-one correspondence, with the exception of the points of g and g', to each of which correspond two points, situated on the opposite sides of the rhombus. However, if we agree to regard these points as identical, two by two, then the correspondence becomes one-one without exception, and *the geometry of the rhombus is completely identical with that of Clifford's Surface.*[1]

§ 11. These representations of the cylinder and of CLIFFORD's Surface show us how, for the case of zero curvature, the investigation of CLIFFORD-KLEIN's forms can be reduced to the determination of suitable Euclidean polygons, eventually degenerating into strips, whose sides are two by two transformable, one into the other, by suitable movements of the plane, their angles being together equal to four right-angles (KLEIN).[2] Then it is only necessary to regard the points of these sides as identical, two by two, to have a representation of the required forms on the ordinary plane.

---

[1] Cf. CLIFFORD loc. cit. Also KLEIN's memoir referred to on p. 211.

[2] Cf. the memoir just named.

It is possible to present, in a similar way, the investigation of CLIFFORD-KLEIN's forms for positive or negative values of the curvature, and the extension of this problem to space.[1]

---

[1] A systematic treatment of CLIFFORD-KLEIN's problem is to be found in KILLING's *Einführung in die Grundlagen der Geometrie*. Bd. I, p. 271—349 (Paderborn, 1893).

Appendix III.

The Non-Euclidean Parallel Construction and other Allied Constructions.

§ 1. The Non-Euclidean Parallel Construction depends upon the correspondence between the right-angled triangle and the quadrilateral with three right angles. Indeed, when this correspondence is known, a number of different constructions are immediately at our disposal. [1]

To express this correspondence we introduce the following notation:

In the right-angled triangle, as usual, a, b are the sides: c is the hypotenuse: λ is the angle opposite a and μ that opposite b. Further the angles of parallelism for a, b are denoted by α and β: and the lines which have λ, μ for angles of parallelism are denoted by l, m. Also two lines, for which the corresponding angles of parallelism are complementary, are distinguished by accents, e. g.:

$$\Pi(a') = \frac{\pi}{2} - \Pi(a), \quad \Pi(l') = \frac{\pi}{2} - \Pi(l).$$

Then with this notation: *To every right-angled triangle* (a, b, c, λ, μ) *there corresponds a quadrilateral with three right-angles, whose fourth angle (acute) is* β, *and whose sides are* c, m', a, l, *taken in order from the corner at which the angle is* β.

The converse of this theorem is also true.

---

[1] Cf. p. 256 of ENGEL's work referred to on p. 84.

The following is one of the constructions, which can be derived from this theorem, for drawing the parallel through A to the line BC (cf. Fig. 77).

Let AB be the perpendicular from A to BC. At A draw the line perpendicular to AB, and from any point C in BC draw the perpendicular CD to this line.

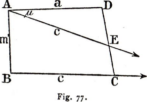

With centre A and radius BC (equal to c) describe a circle cutting CD in E.

Now we have

Fig. 77.

$$\measuredangle\, EAD = \mu,$$

and therefore

$$\measuredangle\, BAE = \frac{\pi}{2} - \mu = \Pi\,(m').$$

But the sides of the quadrilateral are c, m', a, l, taken in order from C.

Therefore AE is parallel to BC.

If a proof of this construction is required without using the trigonometrical forms, one might attempt to show directly that the line AE produced, (simply owing to the equality of BC and AE), does not cut BC produced, and that the two have not a common perpendicular. If this were the case, they would be parallel. Such a proof has not yet been found.

Again, we might prove the truth of the construction using the theorem, that in a prism of triangular section the sum of the three dihedral angles is equal to two right angles[1]: so that for a prism with n angles the sum is $(2n-4)$ right angles. This proof is given in § 2 below.

---

[1] Cf. LOBATSCHEWSKY (ENGEL's translation) p. 172.

Finally, the correspondence stated in the above theorem —only part of which is required for the Parallel Construction of Fig. 78 — can be verified without the use of the geometry of the Non-Euclidean space. This proof is given in § 3.

§ 2. *Direct proof of the Parallel Construction by means of a Prism.*

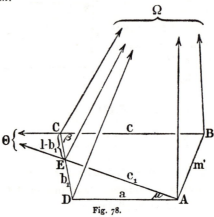

Fig. 78.

Let *ABCD* be a plane quadrilateral in which the angles at $D,"A, B$ are right angles. Let the angle at C be denoted by β, AD by a, DC by l, CB by c, and BA by m'.

At *A* draw the perpendicular $A\Omega$ to the plane of the quadrilateral. Through *B*, *C*, and *D* draw $B\Omega$, $C\Omega$ and $D\Omega$ parallel to $A\Omega$.

Also through *A* draw $A\Theta$ parallel to *BC*, cutting *CD* in E ($ED = b_1$), and let the plane through $A\Omega$ and AE cut $CD\Omega$ in $E\Omega$. From the definition, we have

$$\sphericalangle\,EAD = \frac{\pi}{2} - \Pi\,(m') = \frac{\pi}{2} - \left(\frac{\pi}{2} - \mu\right) = \mu.$$

Further the plane ΩAB is at right angles to *a*, and the plane ΩDA at right angles to *l*, since ΩA and AB are perpendicular to *a*, while ΩD and *a* are perpendicular to *l*.

Also $\quad \measuredangle\, AB\Omega = \measuredangle\, \Theta AB = \dfrac{\pi}{2} - \mu.$

In the prism $\Omega\ (ABCD)$ the faces which meet in ΩA, ΩB, ΩD are perpendicular. Also the four dihedral angles make up four right angles. It follows that the faces of the prism $C\ (DB\Omega)$, which meet along $C\Omega$, are perpendicular. Also it is clear that in $E\ (D\Omega A)$ the faces which meet in EA are perpendicular, while the dihedral angle for the edge CD is the same as for ED (thus equal to α).

We shall now prove the equality of the other dihedral angles in these prisms $C\ (DB\Omega)$ and $E\ (D\Omega A)$—those contained by the faces which meet in CB and AE.

In the first prism this angle is equal to the angle between the planes $ABCD$ and $CB\Omega$. It is thus equal to $\dfrac{\pi}{2} - \mu$, i. e. it is equal to $\measuredangle\, AB\Omega$.

In the second prism, the angle between the planes meeting in $E\Omega$ belongs also to the prism $\Omega\ (ADE)$. In this the angle at ΩD is a right-angle, and that at ΩA is equal to μ. Thus the third angle is equal to $\dfrac{\pi}{2} - \mu$.

Therefore the prisms $C\ (DB\Omega)$ and $E\ (D\Omega A)$ are congruent.

Therefore $\quad \measuredangle\, BC\Omega = \measuredangle\, \Omega EA,$
and the lines which have these angles of parallelism are also equal.

Thus $\quad c = BC$ and $c_1 = AE$
are equal, which was to be proved.

Further it follows that
$$\measuredangle\, DEA = \measuredangle\, DC\Omega;$$
i. e. the angle λ_1, opposite the side a of the triangle, is given by
$$\lambda_1 = \Pi\,(l) = \lambda.$$
Finally $\quad \measuredangle\, DCB = \measuredangle\, DE\Omega;$
i. e. $\quad \beta = \Pi\,(b_1)$, or $b_1 = b.$

Thus the correspondence between the triangle and the quadrilateral is proved.[1]

§ 3. *Proof of the Correspondence by Plane Geometry.*

In the right-angled triangle ABC produce the hypotenuse AB to D, where the perpendicular at D is parallel to CB (cf. Fig. 79).

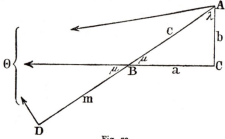

Fig. 79.

Then with the above notation
$$BD = m.$$

Draw through A the parallel to $D\Theta$ and $CB\Theta$.
Then
$$\measuredangle \ CA\Theta = \beta = \Pi (b),$$
and it is also equal to
$$\lambda + \measuredangle \ DA\Theta = \lambda + \Pi (c + m).$$

We thus obtain the first of the six following equations.[2] The third and fifth can be obtained in the same way. The second, fourth, and sixth, come each from the preceding, if we interchange the two sides a and b, and, correspondingly the angles λ and μ.

---

[1] BONOLA: Ist. Lombardo, Rend. (2). T. XXXVII, p. 255—258 (1904). The theorem had already been proved by pure geometrical methods by F. ENGEL: Bull. de la Soc. Phys. Math. de Kasan (2). T. VI (1896); and Bericht d. Kön. Sächs. Ges. d. Wiss., Math.-Phys. Klasse, Bd. L, p. 181—187 (Leipzig, 1898).

[2] Cf. LOBATSCHEWSKY (Engel's translation), p. 15—16, and LIEBMANN, Math. Ann. Bd. LXI, p. 185, (1905).

The table for this case is as follows:

$$\lambda + \Pi\,(c + m) = \beta, \quad \mu + \Pi\,(c + l) = \alpha:$$

$$\lambda + \beta = \Pi\,(c - m), \quad \mu + \alpha = \Pi\,(c - l);$$

$$\Pi\,(b + l) + \Pi\,(m - a) = \frac{1}{2}\pi, \quad \Pi\,(m + a) + \Pi\,(l - b) = \frac{1}{2}\pi.$$

Similar equations can also be obtained for the quadrilateral with three right angles. Some of the sides have to be produced, and the perpendiculars drawn, which are parallel to certain other sides, etc.

If we denote the acute angle of the quadrilateral by β_1, and the sides, counting from it, by c_1, m_1', a_1, and l_1, we obtain the following table:

$$\lambda_1 + \Pi\,(c_1 + m_1) = \beta_1, \quad \gamma_1 + \Pi\,(l_1 + a_1') = \beta_1;$$

$$\lambda_1 + \beta_1 = \Pi\,(c_1 - m_1), \quad \gamma_1 + \beta_1 = \Pi\,(l_1 - a_1');$$

$$\Pi\,(l_1 + b_1) + \Pi\,(m_1 - a_1) = \frac{1}{2}\pi, \quad \Pi\,(c_1 + b_1) + \Pi\,(a_1' - m_1') = \frac{1}{2}\pi.$$

The second, fourth, and sixth formulæ come from interchanging c_1 and m_1', with l_1 and a_1, as in the right-angled triangle.

Let us now imagine a right-angled triangle constructed with the hypotenuse c and the adjacent angle μ: and let the remaining elements be denoted by a, b, λ as above.

In the same way, let a quadrilateral with three right-angles be constructed, in which c is next the acute angle, m' follows c, the remaining elements being a_1, l_1, and β_1.

Then a comparison of the first and third formulæ for the triangle, with the first and third for the quadrilateral, shows that

$$\beta_1 = \beta, \ \lambda_1 = \lambda.$$

The fifth formula of both tables then gives

$$a_1 = a.$$

Hence the theorem is proved.

From the two tables it also follows that to a right-angled triangle with the elements

$$a, b, c, \lambda, \mu,$$

there corresponds a second triangle with the elements

$$a_1 = a, \; b_1 = l', \; c_1 = m, \; \lambda_1 = \frac{\pi}{2} - \beta, \; \mu_1 = \gamma',$$

a result which is of considerable importance in further constructions. But we shall not enter into fuller details.

The possibility of the Non-Euclidean Parallel Construction, with the aid of the ruler and compass, allows us to draw, with the same instruments, the common perpendicular to two lines which are not parallel and do not meet each other (the *non-intersecting lines*); the common parallel to the two lines which bound an angle; and the line which is perpendicular to one of the bounding lines of an acute angle and parallel to the other. We shall now describe, in a few words, how these constructions can be carried out, following the lines laid down by HILBERT.[1]

§ 4. *Construction of the common perpendicular to two non-intersecting straight lines.*

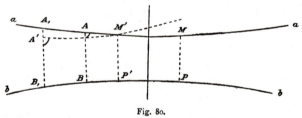

Fig. 80.

Let $a = A_1A$, $b = B_1B$, be two non-intersecting lines; that is, lines which do not meet each other, and are not parallel (cf. Fig. 80).

[1] *Neue Begründung der Bolyai-Lobatschefskyschen Geometrie.* Math. Ann. Bd. 57, p. 137—150 (1903). HILBERT's *Grundlagen der Geometrie*, 2. Aufl., p. 107 et seq.

Let A_1B_1, AB be the perpendiculars drawn from the points A_1, A upon a to the line b, constructed as in ordinary geometry.

If the segments A_1B_1, AB, are equal, the perpendicular to b from the middle point of the segment B_1B is also perpendicular to a; so that, in this case, the construction of the common perpendicular is already effected.

If, on the other hand, the two segments A_1B_1, AB are unequal, let us suppose, e. g., that A_1B_1 is greater than AB.

Then cut off from A_1B_1 the segment $A'B_1$ equal to AB; and through the point A', in the part of the plane in which the segment AB lies, let the ray $A'M'$ be drawn, such that the angle $B_1A'M'$ is equal to the angle which the line a makes with AB (cf. Fig. 80).

The ray $A'M'$ *must cut* the line a in a point M' (cf. HILBERT, loc. cit.). From M' drop the perpendicular $M'F'$ to b, and from the line a, in the direction A_1A, cut off the segment AM equal to $A'M'$.

If the perpendicular MP is now drawn to b, we have a quadrilateral $ABPM$ which is congruent with the quadrilateral $A'B_1F'M'$.

It follows that MP is equal to $M'P'$.

It remains only to draw the perpendicular to b from the middle point of $P'P$ to obtain the common perpendicular to the two lines a and b.

§ 5. *Construction of the common parallel to two straight lines which bound any angle.*

Let $a = AO$, and $b = BO$, be the two lines which contain the angle AOB (cf. Fig. 81). From a and b cut off the equal segments OA and OB; and draw through A the ray b' parallel to the line b, and through B the ray a' parallel to the line a.

Let a_1 and b_1 be the bisectors of the angles contained by the lines ab', and $a'b$.

The two lines a_1b_1 are non-intersecting lines, and their common perpendicular A_1B_1, the construction for which was given in the preceding paragraph, is the common parallel to the lines which bound the angle AOB.

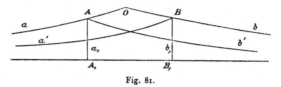

Fig. 81.

Reference should be made to HILBERT's memoir, quoted above, for the proof of this construction.

§ 6. *Construction of the straight line which is perpendicular to one of the lines bounding an acute angle and parallel to the other.*

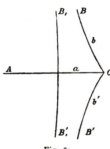

Fig. 82.

Let $a = AO$ and $b = BO$, be the two lines which contain the acute angle AOB; and let the ray $b' = B'O$ be drawn, the image of the line b in a (cf. Fig. 82).

Then, using the preceding construction, let the line B_1B_1' be drawn parallel to the two lines which contain the angle BOB'.

This line, from the symmetry of the figure with respect to a, is perpendicular to OA.

It follows that B_1B_1' is parallel to one of the lines which contain the angle AOB and perpendicular to the other.

§ 7. The constructions given above depend upon metrical considerations. However it is also possible to make use of the fact that to the metrical definitions of perpendicularity and parallelism a projective meaning can be given (§ 79), and that projective geometry is independent of the parallel postulate (§ 80).

Working on these lines, what will be the construction for the parallels through a point A to a given line?

Let the points P_1, P_2, P_3 and P_1', P_2', P_3' be given on g so that the points P_1', P_2', P_3', are all on the same side of P_1, P_2, P_3, and

$$P_1 P_1' = P_2 P_2' = P_3 P_3'.$$

Join AP_1, AP_2, AP_3 and denote these lines by s_1, s_2, and s_3. Similarly let AP_1', AP_2', AP_3' be denoted by s_1', s_2' and s_3'. Then the three pairs of rays through A, determine a projective transformation of the pencil (s) into itself, the double elements of which are obviously the two parallels which we require. These double elements can be constructed by the methods of projective geometry.[1]

The absolute is then determined by five points: i. e., by five pairs of parallels; and so all further problems of metrical geometry are reduced to those of projective geometry.

If we represent (cf. § 84) the LOBATSCHEWSKY-BOLYAI Geometry (e. g., for the Euclidean plane) so that the image of the absolute is a given conic (not reaching infinity), then it has been shown by GROSSMANN[2] that most of the problems for the Non-Euclidean plane can be very beautifully and easily solved by this 'translation'. However we must not forget that this simplicity disappears, if we would pass from the 'translation' back to the 'original text'.

---

[1] Cf. for example, ENRIQUES, *Geometria proiettiva*, (referred to on p. 156) § 73.

[2] GROSSMANN, *Die fundamentalen Konstruktionen der nicht-euklidischen Geometrie*, Programm der Thurgauischen Kantonschule, (Frauenfeld, 1904).

In the Non-Euclidean plane the absolute is inaccessible, and its points are only given by the intersection of pencils of parallels. The points outside of the absolute, while they are accessible in the 'translation', cannot be reached in the 'text' itself. In this case they are pencils of straight lines, which do not meet in a point, but go through the (ideal) pole of a certain line with respect to the absolute.

If, then, we would actually carry out the constructions, difficulties will often arise, such as those we meet in the translation of a foreign language, when we must often substitute for a single adjective a phrase of some length.

———————

Appendix IV.

The Independence of Projective Geometry from Euclid's Postulate.

§ I. *Statement of the Problem.* In the following pages we shall examine more carefully a question to which only passing reference was made in the text (cf. § 80), namely, the validity of Projective Geometry in Non-Euclidean Space, since this question is closely related to the demonstration of the independence of that geometry from the Fifth Postulate.

In elliptic space (cf. § 80) we may assume that the usual projective properties of figures are true, since the postulates of projective geometry are fully verified. Indeed the absence of parallels, or, what amounts to the same thing, the fact that two coplanar lines always intersect, makes the foundation of projectivity in elliptic space simpler than in Euclidean space, which, as is well known, must be first completed by the *points at infinity*.

However in hyperbolic space the matter is more complicated. Here it is not sufficient to account for the absence of the point common to two parallel lines, an exception which destroys the validity of the projective postulate:—*two coplanar lines have a common point.* We must also remove the other exception—the existence of coplanar lines which do not cut each other, and are not parallel (*the non-intersecting lines*). The method, which we shall employ, is the same as that used in dealing with the Euclidean case. We introduce *fictitious points*, regarded as belonging to two coplanar lines which do not meet.

In the following paragraphs, keeping for simplicity to two dimensions only, we show how these fictitious points can be introduced on the hyperbolic plane, and how they enable us to establish the postulates of projective geometry without exception. Naturally no distinction is now made between the *proper points*, that is, the ordinary points, and the *fictitious points*, thus introduced.

§ 2.   *Improper Points and the Complete Projective Plane.* We start with the pencil of lines, that is, the aggregate of the lines of a plane passing through a point. We note that *through any point of the plane, which is not the vertex of the pencil, there passes one, and only one, line of the pencil.*

On the hyperbolic plane, in addition to the pencil, there exist two other systems of lines which enjoy this property, namely;—

(i) *the set of parallels to a line in one direction;*
(ii) *the set of perpendiculars to a line.*

If we extend the meaning of the term, *pencil of lines,* we shall be able to include under it the two systems of lines above mentioned. In that case it is clear *that two arbitrary lines of a plane will determine a pencil, to which they belong.*

If the two lines are *concurrent,* the pencil is formed by the set of lines passing through their common point; if they are *parallel,* by the set of parallels to both, in the same direction; finally, if they are *non-intersecting,* by all the lines which are orthogonal to their common perpendicular. In the first type of pencil (the *proper pencil*), there exists a point common to all its lines, the *vertex of the pencil;* in the two other types (the *improper pencils*), this point is lacking. *We shall now introduce, by convention, a fictitious entity, called an improper point, and regard it as pertaining to all the lines of the pencil.* With this convention, *every pencil has a vertex,* which will be a proper point, or an improper point, according to the different cases. The hyperbolic plane, regarded as the aggregate of all its points, proper and improper, will be called the *complete projective plane*.

§ 3. *The Complete Projective Line.* The improper points are of two kinds. They may be the vertices of pencils of parallels, or the vertices of pencils of non-intersecting lines. The points of the first species are obtained in the same way, and have the same use, as the points at infinity common to two Euclidean parallels. For this reason we shall call them *points at infinity* on the hyperbolic plane, when it is necessary to distinguish them from the others. The points of the second species will be called *ideal points*.

It will be noticed that, while every line has *only one* point at infinity on the Euclidean plane, it has *two* points at infinity on the hyperbolic plane, there being two distinct directions of parallelism for each line. Also that, while the line on the Euclidean plane, with its point at infinity, is closed, the hyperbolic line, regarded as the aggregate of its proper points, and of its two points at infinity, is *open*. The hyperbolic line is closed by associating with it all the ideal points, which are common to it and to all the lines on the plane which do not intersect it.

From this point of view we regard the line as made up of two *segments*, whose common extremities are the two points at infinity of the line. Of these segments, one contains, in addition to its ends, all the proper points of the line; the other all its improper points. The line, regarded as the aggregate of its points, proper and improper, will be called the *complete projective line*.

§ 4. *Combination of Elements.* We assume for the *concrete representation* of a point of the complete projective plane:—

(i) its physical image, if it is a proper point;

(ii) a line which passes through it, and the relative direction of the line, if it is a point at infinity;

(iii) the common perpendicular to all the lines passing through it, if it is an ideal point.

We shall denote a proper point by an ordinary capital letter; an improper point by a Greek capital; and to this we shall add, for an ideal point, the letter which will stand for the representative line of that point. Thus a point at infinity will be denoted, e. g., by Ω, while the ideal point, through which all lines perpendicular to the line o pass, will be denoted by Ω_o.

On this understanding, if we make no distinction between proper points and improper points, not only can we affirm the unconditional validity of the projective postulate: *two arbitrary lines have a common point:* but we can also construct this point, understanding by this construction the process of obtaining its concrete representation. In fact, if the lines meet, in the ordinary sense of the term, or are parallel, the point can be at once obtained. If they are non-intersecting, it is sufficient to draw their common perpendicular, according to the rule obtained in Appendix III § 4.

On the other hand, we are not able to say that the second postulate of projective geometry—*two points determine a line*—and the corresponding constructions, are valid unconditionally. In fact no line passes through the ideal point Ω_o and through the point at infinity Ω on the line o, since there is no line which is at the same time parallel and perpendicular to a line o.

Before indicating how we can remove this and other exceptions to the principle that a line can be determined by a pair of points, we shall enumerate all the cases in which two points fix a line, and the corresponding constructions:—

a) *Two proper points.* The line is constructed as usual.

b) *A proper point* [O] *and a point at infinity* [Ω]. The line OΩ is constructed by drawing the parallel through O to the line which contains Ω, in the direction corresponding to Ω. (Appendix III).

(c) *A proper point* [O] *and an ideal point* [Γ_c]. The line OΓ_c is constructed by dropping the perpendicular from O to the line *c*.

(d) *Two points at infinity* [Ω, Ω']. The line ΩΩ' is the common parallel to the two lines bounding an angle, the construction for which is given in Appendix III § 5.

(e) *An ideal point* [Γ_c] *and a point at infinity* [Ω], *not lying on the representative line c of the ideal point*. The line ΩΓ_c is the line which is parallel to the direction given by Ω and perpendicular to *c*. The construction is given in Appendix III § 6.

(f) *Two ideal points* [Γ_c, Γ_c'], *whose representative lines c, c' do not intersect*. The line $\Gamma_c\Gamma_c'$, is constructed by drawing the common perpendicular to *c* and *c'* (Appendix III § 4).

The pairs of points which do not determine a line are as follows:—

(i) an ideal point and a point at infinity, lying on the representative line of the ideal point;

(ii) two ideal points, whose representative lines are parallel, or meet in a proper point.

§ 5. *Improper Lines*. To remove the exceptions mentioned above in (i) and (ii), new entities must be introduced. These we shall call *improper lines*, to distinguish them from the ordinary or *proper lines*.

These improper lines are of two types:—

(i) If Ω is a point at infinity, every line of the pencil Ω is the representative entity of an ideal point. The locus of these ideal points, together with the point Ω, is an improper line of the first type, or *line at infinity*. It will be denoted by ω.

(ii) If A is a proper point, every line passing through A is the representative entity of an ideal point. The locus of these ideal points is an improper line of the second type, or *ideal line.* It will be denoted by α_A. The proper point A can be taken as representative of the ideal line α_A.

These definitions of the terms *line at infinity* and *ideal line* allow us to state that two points, which do not belong to a proper line, determine either a line at infinity, or an ideal line. Hence, dropping the distinction between proper and improper elements, the projective postulate—*two points determine a line*—is universally true.

We must now show that, with the addition of the improper lines, any two lines have a common point. The various cases in which the two lines are proper have been already discussed (§ 4). There remain to be examined the cases in which at least one of the lines is improper.

(i) Let r be a proper line and ω an improper line, passing through the point Ω at infinity. The point ωr is the ideal point, which has the line passing through Ω and perpendicular to r for representative line.

(ii) Let r be a proper line and α_A an ideal line. The point $r\alpha_A$ is the ideal point, which has the line passing through A and perpendicular to r for its representative line.

(iii) Let ω and ω' be two lines at infinity, to which belong the points Ω and Ω' respectively. The point $\omega\omega'$ is the ideal point, whose representative line is the line joining the points Ω and Ω'.

(iv) Let α_A, β_B be two ideal lines. The point $\alpha_A\beta_B$ is the ideal point, whose representative line is the line joining A and B.

(v) Let ω and α_A be a line at infinity and an ideal line. The point $\omega\alpha_A$ is the ideal point, whose representative line is the line joining A to Ω.

Thus we have demonstrated that the two fundamental

postulates of projective plane geometry hold on the hyperbolic plane.

§ 6. *Complete Projective Space and the Validity of Projective Geometry in the Hyperbolic Space.* We can introduce improper points, lines and planes, into the Hyperbolic Space by the same method which has been followed in the preceding paragraphs. We can then extend the fundamental propositions of projective geometry to the *complete projective space.* Thereafter, following the lines laid down by STAUDT, all the important projective properties of figures can be demonstrated. Thus the validity of projective geometry in the LOBATSCHEWSKY-BOLYAI Space is established.

§ 7. *Independence of Projective Geometry from the Fifth Postulate.* Let us suppose that in a connected argument, founded on the group of postulates $A, B \ldots \ldots, H$, the *only* hypotheses which can be used are $I_1, I_2 \ldots \ldots, I_n$. Also that from the fundamental postulates and any one whatever of the $I's$, a certain proposition M can be derived. Then we may say that M is independent of the $I's$.

It is precisely in this way that the independence of projective geometry from the Fifth Postulate is proved, since we have shown that it can be built up, starting from the group of postulates common to the three systems of geometry, and then adding to them any one of the hypotheses on parallels.

The demonstration of the independence of M from any one of the $I's$, founded on the deduction (cf. § 59)

$$\left\{A, B, \ldots H, I_r\right\} \supset M_{(r\, =\, 1,\, 2,\, \ldots\, n)}$$

may be called *indirect,* reserving the term *direct demonstration for that which shows that it is possible to obtain M without introducing any of the I's at all.* Such a possibility, from the theoretical point of view, is to be expected, since the preceding relations show that neither any single I, nor any group of them, is necessary to obtain M. If we wish to give a demonstration of the type

$$\{A, B, \ldots H\} \supset M,$$

in which the I's do not appear at all, we may meet difficulties not always easily overcome, difficulties depending on the nature of the question, and on the methods we may adopt to solve it. So far as regards the independence of projective geometry from the Fifth Postulate, we possess two interesting types of direct proofs, founded on two different orders of ideas. One employs the method of analysis: the other that of synthesis. We shall now briefly describe the views on which they are founded.

§ 8. *Beltrami's Direct Demonstration of the Independence of Projective Geometry from the Fifth Postulate.* The demonstration implicitly contained in BELTRAMI's '*Saggio*' of 1868 must be placed first in chronological order. Referring to the '*Saggio*', *let us suppose that between the points of a surface F, (or of a suitably limited region of the surface), and the points of an ordinary plane area, there can be established a one-one correspondence, such that the geodesics of the former are represented by the straight lines of the latter.* Then, to the projective properties of plane figures, which express the collinearity of certain points, the concurrence of certain lines, etc., correspond similar properties of the corresponding figures on the surface, which are deduced from the first, by simply changing the words *plane* and *line* into *surface* and *geodesic*. If all this is possible, we should naturally say that the projective properties of the corresponding plane area are valid on the surface F; or, more simply, that the ordinary projectivity of the plane holds upon the surface. We shall now put this result in an analytical form.

Let u and v be the (curvilinear) coordinates of a point

on F, and x and y those of the representative point on the plane. The correspondence between the points (u, v) and (x, y) will be expressed analytically by putting

$$\left. \begin{array}{l} u = f\,(x,\,y) \\ v = \varphi\,(x,\,y) \end{array} \right\}, \qquad (1)$$

where f and φ are suitable functions.

To the equation

$$\psi\,(u,\,v) = o$$

of a geodesic on F, let us now apply the transformation (1).

We must obtain a *linear equation* in x, y, since, by our hypothesis, the geodesics of F are represented by straight lines on the plane.

But the equations (1) can also be interpreted as formulæ giving a *transformation of coordinates* on F. We can therefore conclude that:—*If, by a suitable choice of a system of curvilinear coordinates on the surface F, the geodesics of that surface can be represented by linear equations, the ordinary projective geometry is valid on the surface.*

Now BELTRAMI has shown in his '*Saggio*' that on surfaces of constant curvature it is always possible to choose a system of coordinates (u, v), for which the general integral of the differential equation of the geodesics takes the form

$$ax + by + c = o.$$

Hence, from what has been said above, it follows that:—

Plane projective geometry is valid on the surfaces of constant curvature with respect to their geodesics.

But, according to the value of the curvature, the geometry of these surfaces coincides with that of the Euclidean plane, or of the Non-Euclidean planes.

It follows that:—

The method of Beltrami, applied to a plane on which are valid the metrical concepts common to the three geometries, leads to the foundation of plane projective geometry without the assumption of any hypothesis on parallels.

This result and the argument we have employed in obtaining it are easily extended to space. BELTRAMI's memoir referring to this is the *Teoria fondamentale degli spazii di curvatura costante*, quoted in the note to § 75.

§ 9. *Klein's Direct Demonstration of the Independence of Projective Geometry from the Fifth Postulate.* The method indicated above is not the only one which will serve our purpose. In fact, we might be asked if we could not construct projective geometry independently of any metrical consideration; that is, starting from the notions of point, line, plane, and from the axioms of connection and order, and the principle of continuity.[1] In 1871 KLEIN was convinced of the possibility of such a foundation, from the consideration of the method followed by STAUDT in the construction of his geometrical system. There remained one difficulty, relative to the improper points. STAUDT, following PONCELET, makes them to depend on the ordinary theory of parallels. To escape the various exceptions to the statement that two coplanar lines have a common point, due to the omission of the Euclidean hypothesis, KLEIN *proposed to construct projective geometry in a limited (and convex) region of space,* such, e. g., as that of the points inside a tetrahedron. With reference to such a region, for the end he has in view, every point on the faces of, or external to, the tetrahedron must be considered as non-existent. Also we *must give the name of line and plane only to the portions of the line and plane belonging to the region considered.* Then the graphical postulates of connection, order, etc., which are supposed true in the whole

---

[1] For this nomenclature for the Axioms, cf. TOWNSEND's translation of HILBERT's *Foundations of Geometry*, p. 1 (Open Court Publishing Co. 1902).

of space, are verified in the interior of the tetrahedron. Thus to construct projective geometry in this region, it is necessary, with suitable conventions, that the propositions on the concurrence of lines, etc. should hold without exception. These are not always true, when the word point means simply point inside the tetrahedron.

KLEIN showed briefly, while various later writers discussed the question more fully, how the space inside the tetrahedron can be completed by fictitious entities, called ideal points, lines and planes, so that when no distinction is made between the proper entities (inside the tetrahedron) and the ideal entities, the graphical properties of space, on which all projective geometry is constructed, are completely verified.

From this there readily follows the independence of projective geometry from EUCLID's Fifth Postulate.

Appendix V.

The Impossibility of Proving Euclid's Parallel Postulate.[1]

An Elementary Demonstration of this Impossibility founded upon the Properties of the System of Circles orthogonal to a Fixed Circle.

§ 1. In the concluding article (§ 94) various arguments are mentioned, any one of which establishes the independence of EUCLID's Parallel Postulate from the other assumptions on which Euclidean Geometry is based. One of these has been discussed in greater detail in Appendix IV. In the articles which follow there will be found another and a more elementary proof that the BOLYAI-LOBATSCHEWSKY system of Non-Euclidean Geometry cannot lead to any contradictory results, and that it is therefore impossible to prove EUCLID's Postulate or any of its equivalents. This proof depends, for solid geometry, upon the properties of the system of spheres all orthogonal to a fixed sphere, while for plane geometry the system of circles all orthogonal to a fixed circle is taken. In the course of the discussion many of the results of Hyperbolic Geometry are deduced from the properties of this system of circles.

---

[1] This Appendix, added to the English translation, is based upon WELLSTEIN's work, referred to on p. 180, and the following paper by CARSLAW; *'The Bolyai-Lobatschewsky Non-Euclidean Geometry: an Elementary Interpretation of this Geometry and some Results which follow from this Interpretation,* Proc. Edin. Math. Soc. Vol. XXVIII, p. 95 (1910).

Cf. also: J. WELLSTEIN, *Zusammenhang zwischen zwei euklidischen Bildern der nichteuklidischen Geometrie.* Archiv der Math. u. Physik (3). XVII, p. 195 (1910).

The System of Circles passing through a fixed Point.

§ 2. We shall examine first of all the representation of ordinary Euclidean Geometry by the geometry of the system of spheres all passing through a fixed point. In plane geometry this reduces to the system of circles through a fixed point, and we shall begin with that case.

Since the system of circles through a point O is the inverse of the system of straight lines lying in the plane, to every circle there corresponds a straight line, and the circles intersect at the same angle as the corresponding lines. The properties of the set of circles could be established from the knowledge of the geometry of the straight lines, and every proposition concerning points and straight lines in the one geometry could at once be interpreted as a proposition concerning points and circles in the other.

There is another way in which the geometry of these circles can be established independently. We shall first describe this method, 'and we shall then see that from this interpretation of the Euclidean Geometry we can easily pass to a corresponding representation of the Non-Euclidean Geometry.

§ 3. *Ideal Lines.*

It will be convenient to speak ot the plane of the straight lines and the plane of the circles, as two separate planes. We have seen that to every straight line in the plane of the straight lines, there corresponds a circle in the plane of the circles. We shall call these circles *Ideal Lines*. The *Ideal Points* will be the same as ordinary points, except that the point O will be excluded from the domain of the Ideal Points.

On this understanding we can say that *Any two different Ideal Points, A, B, determine the Ideal Line AB*; just as, in Euclidean Geometry, any two different points A, B determine the straight line AB.

As the angle between the circles in the one plane is equal to the angle between the corresponding straight lines in the other, we define *the angle between two Ideal Lines as the angle between the corresponding straight lines.* Thus we can speak of Ideal Lines being perpendicular to each other, or cutting at any angle.

§ 4.   *Ideal Parallel Lines.*

Let BC (cf. Fig. 83) be any straight line and A a point not lying upon it.

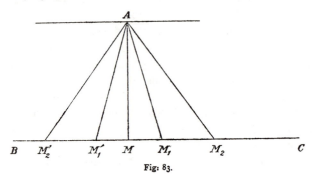

Fig: 83.

Let AM be the perpendicular to BC, and AM_1, AM_2, AM_3, ... different positions of the line AM, as it revolves from the perpendicular position through two right angles.

The lines begin by cutting BC on the one side of M, and there is one line separating the lines which intersect BC on the one side, from those which intersect it on the other. This line is the parallel through A to BC.

In the corresponding figure for the Ideal Lines (cf. Fig. 84), we have the Ideal Line through A perpendicular to the Ideal Line BC; and the circle which passes through A, and touches the circle OBC at O, separates the circles through A, which cut BC on the one side of M, from those which cut it on the other.

We are thus led to define *Parallel Ideal Lines* as follows:

The Ideal Line through any point parallel to a given Ideal Line is the circle of the system which touches at O the circle coinciding with the given line and also passes through the given point.

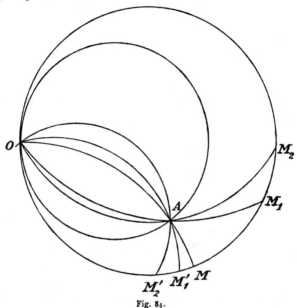

Fig. 84.

Thus any two circles of the system which touch each other at O will be Ideal Parallel Lines. Two Ideal Lines, which are each parallel to a third Ideal Line, are parallel to each other, etc.

§ 5. *Ideal Lengths.*

Since EUCLID's Parallel Postulate is equivalent to the assumption that one, and only one, straight line can be drawn through a point parallel to another straight line, and since this postulate is obviously satisfied by the Ideal Line, in the geometry of these lines, EUCLID's Theory of Parallels will be true.

But such a geometry will require a measurement of length. We must now define what is meant by the *Ideal Length of an Ideal Segment*. In other words we must define the *Ideal Distance* between two points. It is clear that if the two geometries are to be identical two Ideal Segments must be regarded as of equal length, when the corresponding rectilinear segments are equal. We thus define the *Ideal Length of an Ideal Segment as the length of the rectilinear segment to which it corresponds*.

It will be seen that the Ideal Distance between two points A, B is such that, if C is any other point on the segment,

'distance' AB = 'distance' AC + 'distance' CB.

The other requisite for 'distance' is that it is unaltered by displacement, and when we come to define *Ideal Displacement* we shall have to make sure that this condition is also satisfied.

It is clear that on this understanding the Ideal Length of an Ideal Line is infinite. If we take 'equal' steps along the Ideal Line BC from the foot of the perpendicular (cf. Fig. 84) the actual lengths of the arcs MM_1, M_1M_2, etc...., the Ideal Lengths of which are equal, become gradually smaller and smaller, as we proceed along the line towards O. It will take an infinite number of such steps to reach O, just as it will take an infinite number of steps along BC from M (cf. Fig. 83) to reach the point at which BC is met by the parallel through A. We have already seen that the domain of Ideal Points contains all the points of the plane except O. This was required so that the Ideal Line might always be determined by two different points. It is also needed for the idea of 'between-ness'. On the straight line AB we can say that C lies between line A and B if, as we proceed along

AB from *A* to *B*, we pass through *C*. On the Ideal Line *AB* (cf. Fig. 85) the points C_1 and C_2 would both lie between *A* and *B*, unless the point *O* were excluded. In other words this convention must be made so that the Axioms of Order [1] may appear in the geometry of the Ideal Points and Lines.

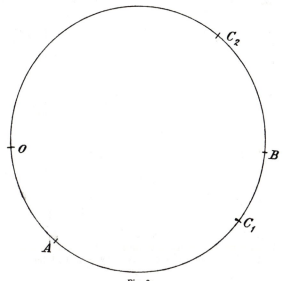

Fig 85.

On this understanding, and still speaking of plane geo-metry, we can say that *two Ideal Lines are parallel when they do not meet, however far they are produced.*

To obtain an expression for the Ideal Length of an Ideal Segment we may take the radius of inversion—*k*—to be unity.

Consider the segment *AB* and the rectilinear segment αβ to which it corresponds. Then we have (Fig. 86)

$$\frac{\alpha\beta}{AB} = \frac{O\beta}{OA} = \frac{O\beta . OB}{OA . OB} = \frac{k^2}{OA . OB}.$$

[1] See Note on p. 236.

Hence we define *the Ideal Length of the segment AB* as

$$\frac{AB}{OA \cdot OB}$$

We shall now show that *the Ideal Length of an Ideal Segment is unaltered by inversion with regard to any circle of the system.*

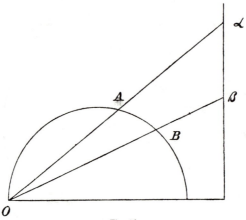

Fig. 86.

Let OD be any circle of the system and let C be its centre (Fig. 87).

Then inversion changes an Ideal Line into an Ideal Line.

Let the Ideal Segment AB invert into the Ideal Segment $A'B'$. These two Ideal Lines intersect at the point D, where the circle of inversion meets AB.

Then

$$\frac{\text{the Ideal Length of } AD}{\text{the Ideal Length of } A'D} = \frac{AD}{OA \cdot OD} \bigg/ \frac{A'D}{OA' \cdot OD}$$

$$= \frac{AD}{A'D} \cdot \frac{OA}{OA'}$$

But from the triangles CAD, $CA'D$ and OAC, $OA'C$, we find

$$\frac{AD}{A'D} = \frac{CA}{CD} = \frac{CA}{CO} = \frac{AO}{A'O}.$$

Thus the Ideal Length of AD = the Ideal Length of $A'D$. Similarly we find BD and $B'D$ have the same Ideal Length, and therefore AB and $A'B'$ have the same Ideal Length.

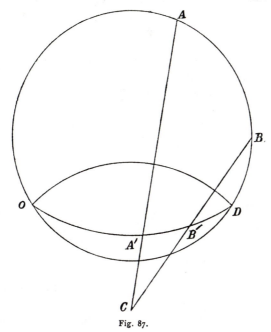

Fig. 87.

§ 6. *Ideal Displacements.*

The length of a segment must be unaltered by displacement. This leads us to consider the definition of *Ideal Displacement*. Any displacement may be produced by repeated applications of reflection; that is, by taking the image of the figure in a line (or in a plane, in the case of solid geometry). For example, to translate the segment AB (cf. Fig. 88) into another position on the same straight line, we may reflect the figure, first about a line perpendicular to and bisecting BB', and then another reflection about the middle point of $A'B'$ would bring the ends into their former positions relative to each other. Also to move the segment AB into

Fig. 88.

the position $A'B'$ (cf. Fig. 89) we can first take the image of AB in the line bisecting the angle between AB and $A'B'$, and then translate the segment along $A'B'$ to its final position.

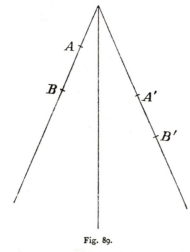

Fig. 89.

We proceed to show that inversion about any circle of the system is equivalent to reflection of the Ideal Points and Lines in the Ideal Line which coincides with the circle of inversion.

Let C (Fig. 90) be the centre of any circle of the system, and let A' be the inverse of any point A with regard to this circle. Then the circle OAA' is orthogonal to the circle of inversion.

In other words, such inversion changes any point A into a point A' on the Ideal Line perpendicular to the circle of inversion. Also the Ideal Line AA' is 'bisected' by that circle at M, since the Ideal Segment AM inverts into the segment $A'M$, and Ideal Lengths are unaltered by such inversion.

Again let AB be any Ideal Segment, and by inversion

with regard to any circle of the system let it take up the position $A'B'$ (Fig. 87). We have seen that the Ideal Length of the segment is unaltered: and it is clear that the two segments, when produced, meet on the circle of inversion, and make equal angles with it. Also the Ideal Lines AA'

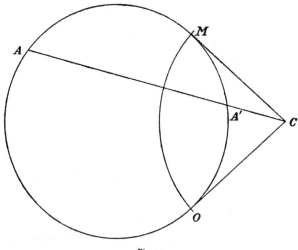

Fig. 90.

and BB' are perpendicular to, and 'bisected' by, the Ideal Line with which the circle of inversion coincides.

Such an inversion is, therefore, the same as reflection, and translation will occur as a special case of the above, when the circle of inversion is orthogonal to the given Ideal Line.

We thus define Ideal Reflection in an Ideal Line as inversion with this line as the circle of inversion.

It is unnecessary to say more about *Ideal Displacements* than that they will be the result of Ideal Reflection.

With these definitions it is now possible to 'translate' every proposition in the ordinary plane geometry into a corresponding proposition in this Ideal Geometry. We have only to use the words Ideal Points, Lines, Parallels, etc., instead of the ordinary points, lines, parallels, etc. The argument employed in proving a theorem, or the construction used in solving a problem, will be applicable, word for word, in the one geometry as well as in the other, for the elements involved satisfy the same laws. This is the 'dictionary' method so frequently adopted in the previous pages of this book.

§ 7. *Extension to Solid Geometry. The System of Spheres passing through a fixed point.*

These methods may be extended to solid geometry. In this case the inversion of the system of points, lines, and planes gives rise to the system of points, circles intersecting in the centre of inversion, and spheres also intersecting in that point. The geometry of this system of spheres could be derived from that of the system of points, lines and planes, by interpreting each proposition in terms of the inverse figures. For our purpose it is better to regard it as derived from the former by the invention of the terms: Ideal Point, Ideal Line, Ideal Plane, Ideal Length and Ideal Displacement.

The *Ideal Point* is the same as the ordinary point, but the point O is excluded from the domain of Ideal Points.

The *Ideal Line* through two Ideal Points is the circle of the system which passes through these two points.

The *Ideal Plane* through three Ideal Points, not on an Ideal Line, is the sphere of the system which passes through these three points.

Thus the plane geometry, discussed in the preceding articles, is a special case of this plane geometry.

Ideal Parallel Lines are defined as before. The line through A parallel to BC is the circle of the system, lying on the sphere through O, A, B, and C, which touches the circle given by the Ideal Line BC at O and passes through A.

It is clear that an Ideal Line is determined by two points, as a straight line is determined by two points. An Ideal Plane is determined by three points, not on an Ideal Line, as an ordinary plane is determined by three points, not on a straight line. If two points of an Ideal Line lie on an Ideal Plane, all the points of the line do so: just as if two points of a straight line lie on a plane, all its points do so. The intersection of two Ideal Planes is an Ideal Line; just as the intersection of two ordinary planes is a straight line.

The measurement of angles in the two spaces is the same.

For the measurement of length we adopt the same definition of *Ideal Length* as in the case of two dimensions. The Ideal Length of an Ideal Segment is the length of the rectilinear segment to which it corresponds. To these definitions it only remains to add that of *Ideal Displacement*. As in the two dimensional case, this is reached by means of *Ideal Reflection:* and it can easily be shown that *if the system of Ideal Points, Lines and Planes is inverted with regard to one of its spheres, the result is equivalent to a reflection of the system in this Ideal Plane.*

This Ideal Geometry is identical with the ordinary Euclidean Geometry. Its elements satisfy the same laws: every proposition valid in the one is also valid in the other: and from the results of Euclidean Geometry those of the Ideal Geometry can be inferred.

In the articles that follow we shall establish an Ideal Geometry whose elements satisfy the axioms upon which the Non-Euclidean Geometry of BOLYAI-LOBATSCHEWSKY is based. The points, lines and planes of this geometry will be figures of the Euclidean Geometry, and from the known properties of these figures, we could state what the corresponding theorems of this Non-Euclidean Geometry would be. Also from some of its constructions, the Non-Euclidean constructions could be obtained. This process would be the converse of that referred to in dealing with the Ideal Geometry of the preceding articles: since, in that case, we obtained the theorems of the Ideal Geometry from the corresponding Euclidean theorems.

The Geometry of the System of Circles Orthogonal to a Fixed Circle.

§ 8. *Ideal Points, Ideal Lines and Ideal Parallels.*

In the Ideal Geometry discussed in the previous articles, the Ideal Point was the same as the ordinary point, and the Ideal Lines and Planes had so far the characteristics of straight lines and planes that they were lines and surfaces respectively. Geometries can be constructed in which the Ideal Points, Lines and Planes are quite removed from ordinary points, lines, and planes: so that the Ideal Points no longer have the characteristic of having no parts: and the Ideal Lines no longer boast only length, etc. What is required in each geometry is that the entities concerned satisfy the axioms which form the foundations of geometry. If they satisfy the axioms of Euclidean Geometry, the arguments, which lead to the theorems of that geometry, will give corresponding theorems in the Ideal Geometry: and if they satisfy the axioms of any of the Non-Euclidean Geometries, the arguments, which lead to theorems in that Non-Euclidean Geometry, will lead equally to theorems in the corresponding Ideal Geometry.

We proceed to discuss the geometry of the system of circles orthogonal to a fixed circle.

Let the fundamental circle be of radius k and centre O.

Let A', A'' be any two inverse points, A' being inside the circle. *Every such pair of points* (A', A''), *is an Ideal Point* (A) *of the Ideal Geometry with which we shall now deal.*

If two such pairs of points are given—that is, two Ideal Points (A, B), (Fig. 92)—these determine a circle which is orthogonal to the fundamental circle. *Every such circle is an Ideal Line of this Ideal Geometry.*

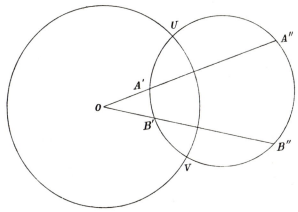

Fig. 91.

Hence any two different Ideal Points determine an Ideal Line. In the case of the system of circles passing through a fixed point O, this point O was excluded from the domain of the Ideal Points. In this system of circles all orthogonal to the fundamental circle, the coincident pairs of points lying on the circumference of that circle are excluded from the domain of the Ideal Points.

We define the angle between two Ideal Lines as the angle between the circles which coincide with these lines.

We have now to consider in what way it will be proper to define *Parallel Ideal Lines.*

Let AM be the Ideal Line through A, perpendicular to the Ideal Line BC; in other words, the circle of the system passing through A', A'', and orthogonal to the circle through B', B'', C' and C'' (cf. Fig. 92).

Imagine AM to rotate about A so that those Ideal Lines through A cut the Ideal Line BC at a gradually decreasing angle. The circles through A which touch the given

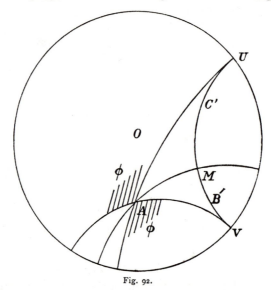

Fig. 92.

circle BC at the points U, V, where it meets the fundamental circle, are Ideal Lines of the system. They separate the lines of the pencil of Ideal Lines through A, which cut the Ideal Line BC, from those which do not cut that line. All the lines in the angle φ, shaded in the figure, do not cut the line BC; all those in the angle ψ, not shaded, do cut this line. This property is exactly what is assumed in the Parallel Postulate upon which the Non-Euclidean Geometry of Bolyai-Lobatschewsky is based. We therefore are led to define Parallel Ideal Lines in this Plane Ideal Geometry as follows:

The Ideal Lines through an Ideal Point parallel to a given Ideal Line are the two circles of the system passing through the given point, which touch the circle with which the given line coincides at the points where it meets the fundamental circle.

Thus we have in this Ideal Geometry two parallels through a point to a given line: a right-handed parallel, and a left-handed parallel: and these separate the lines of the pencil which intersect the given line from those which do not intersect it.

Some Theorems of this Non-Euclidean Geometry.

§ 9. At this stage we can say that any of the theorems of the BOLYAI-LOBATSCHEWSKY Non-Euclidean Geometry, involving angle properties only, will hold in this Ideal Geometry and vice versa. Those involving lengths we cannot yet discuss, as we have not yet defined *Ideal Lengths*. For example, it is obvious that there are triangles in which all the angles are zero (cf. Fig. 93). The sides of such triangles are parallel in pairs. Thus the sum of the angles of an Ideal Triangle is certainly not always equal to two right angles. We can prove that this sum is always less than two right angles by a simple application of inversion, as follows:

Let C_1, C_2, C_3 be three circles of the system, forming an Ideal Triangle. Invert these circles from the point of intersection I of C_1 and C_2, which lies inside the fundamental circle. Then C_1 and C_2 become two straight lines C_1' and C_2' through I. Also the fundamental circle C inverts into a circle C' cutting C_1' and C_2' at right-angles, so that its centre is I. Again, the circle C_3 inverts into a circle C_3', cutting C' at right-angles. Hence its centre lies outside C'. We thus obtain a 'triangle', in which the sum of the angles is less than two right-angles, and since these angles are equal to the angles of the Ideal Triangle, this result holds also for the Ideal Triangle.

Finally, it can be shown that there is always one, and only one, circle of the system cutting two non-intersecting circles of the system at right-angles. In other words, two

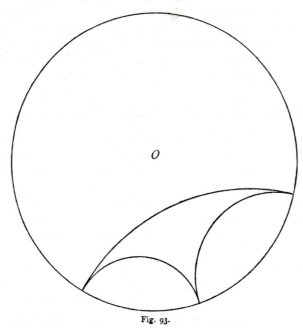

Fig. 93.

non-intersecting Ideal Lines have a common perpendicular. All these results must be true in the Hyperbolic Geometry.

§ 10. *Ideal Lengths and Ideal Displacements.*

Before we can proceed to the discussion of the metrical properties of this geometry, we must define the *Ideal Length of an Ideal Segment*. It is clear that this must be such that it will be unaltered, if we take the points A'', B'', as defining the segment AB, instead of the points A', B'. It must make the complete line infinite in length. It must satisfy the distributive law 'distance' $AB =$ 'distance' $AC +$ 'distance' CB, if C is any other point on the segment AB, and it must also remain unaltered by *Ideal Displacement*.

We define the Ideal Length of any segment AB as

$$\log \left(\frac{A'V}{A'U} \middle| \frac{B'V}{B'U} \right)$$

where U, V are the points where the Ideal Line AB meets the fundamental circle (cf. Fig. 91).

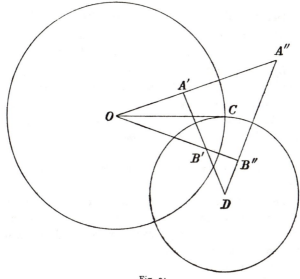

Fig. 94.

This expression obviously involves the Anharmonic Ratio of the points $UABV$. It will be seen that this definition satisfies the first three of the conditions named above. It remains for us to examine what must represent displacement in this Ideal Geometry.

Let us consider what is the effect of inversion with regard to a circle of the system upon the system of Ideal Points and Lines.

Let $A'A''$ be any Ideal Point A (cf. Fig. 94). Let the

circle of inversion meet the fundamental circle in C, and let D be its centre. Let A', A'' invert into B', B''. Since the circle $A'A''C$ touches the circle of inversion at C, its inverse also touches that circle at C. But a circle passes through A', A'', B' and B'', and the radical axes of the three circles

$$A'A''C,\ B'B''C,\ A'A''B'B''$$

are concurrent.

Hence $B'B''$ passes through O, and $OB' \cdot OB'' = OC^2$.

Therefore *inversion with regard to any circle of the system changes an Ideal Point into an Ideal Point.*

But it is clear that the circle $A'A''B'B''$ is orthogonal to the fundamental circle, and also to the circle of inversion.

Thus *the Ideal Line joining the Ideal Point A and the Ideal Point B, into which it is changed by this inversion, is perpendicular to the Ideal Line coinciding with the circle of inversion.*

We shall now prove that it is 'bisected' by that Ideal Line.

Let the circle through AB meet the circle of inversion at M, and the fundamental circle in U and V. It is clear that U and V are inverse points with regard to the circle of inversion [cf. Fig. 95].

Then we have:

$$\frac{B'V}{A'U} = \frac{CV}{CA'},$$

$$\text{and } \frac{A'V}{B'U} = \frac{CV}{CB'}.$$

Thus

$$\frac{A'V}{A'U}\ \frac{B'V}{B'U} = \frac{CV^2}{CA' \cdot CB'} = \frac{CV^2}{CM^2} = \left(\frac{M'V}{M'U}\right)^2.$$

Therefore

$$\frac{A'V}{A'U} \bigg/ \frac{M'V}{M'U} = \frac{M'V}{M'U} \bigg/ \frac{B'V}{B'U}.$$

Hence the Ideal Length of *AM* is equal to the Ideal
Length of *MB*.

Thus we have the following result:

Inversion with regard to a circle of the system changes
any Ideal Point A into an Ideal Point B, such that the Ideal
Line AB is perpendicular to, and 'bisected' by, the Ideal Line
coinciding with the circle of inversion.

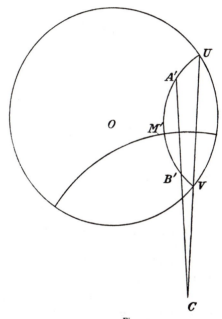

Fig. 95

In other words, *inversion with regard to such a circle*
causes any Ideal Point A to take the position of its image in
the corresponding Ideal Line.

We proceed to examine what effect such inversion has
upon an Ideal Line.

Since a circle, orthogonal to the fundamental circle,

inverts into a circle also orthogonal to the fundamental circle, any Ideal Line AB inverts into another Ideal Line ab, passing through the point M, where AB meets the circle of inversion (cf. Fig. 96). Also the points U, V invert into the

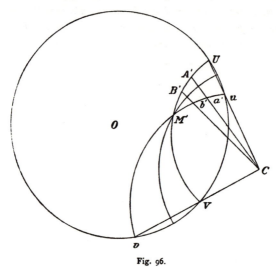

Fig. 96.

points u and v on the fundamental circle; and the lines AB and ab are equally inclined to the circle of inversion.

It is easy to show that the Ideal Lengths of AM and BM are equal to those of aM and bM respectively, and it follows that the Ideal Length of the segment AB is unaltered by this inversion. Also we have seen that Aa and Bb are perpendicular to, and 'bisected' by, the Ideal Line coinciding with this circle.

It follows from these results that inversion with regard to any circle of the system has the same effect upon an Ideal Segment as reflection in the corresponding Ideal Line.

We are thus again able to define Ideal Reflection in any Ideal Line as the inversion of the system of Ideal Points and

Lines with regard to the circle which coincides with this Ideal Line.

It is unnecessary to define *Ideal Displacements*, as any displacement can be obtained by a series of reflections and any Ideal Displacement by a series of Ideal Reflections.

We notice that the definition of the Ideal Length of any Segment fixes the *Ideal Unit of Length*. We may take this on one of the diameters of the fundamental circle, since these lines are also Ideal Lines of the system. Let it be the segment OP (Fig. 97).

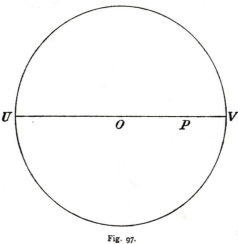

Fig. 97.

Then we must have

$$\log \left(\frac{OV}{OU} \,\middle|\, \frac{PV}{PU} \right) = 1 ;$$

i. e.
$$\log \frac{PU}{PV} = 1.$$

Therefore
$$\frac{PU}{PV} = e,$$

and the point P divides the diameter in the ratio $e : 1$.

The Unit Segment is thus fixed for any position in the

domain of the Ideal Points, since the segment OP can be 'moved' so that one of its ends coincides with any given Ideal Point.

A different expression for the Ideal Length

$$k \log \left(\frac{AV}{AU} \, \bigg| \, \frac{BV}{BU} \right)$$

would simply mean an alteration in the unit, and taking logarithms to any other base than e would have the same effect.

§ 11. *Some further Theorems in this Non-Euclidean Geometry.*

We are now in a position to establish some further theorems of the Hyperbolic Geometry using the metrical properties of this Ideal Geometry.

In the first place we can state that Similar Triangles are impossible in this geometry.

We also see that Parallel Ideal Lines are asymptotic; that is, these lines continually approach each other and the distance between them tends to zero.

Further, it is obvious that as the point A moves away along the perpendicular MA to the line BC (cf. Fig. 92), the angle of parallelism diminishes from $\dfrac{\pi}{2}$ to zero in the limit.

Again, we can prove from the Ideal Geometry that the Angle of Parallelism $\Pi(p)$, corresponding to a segment p, is given by

$$\tan \frac{\Pi(p)}{2} = e^{-p}$$

Consider an Ideal Line and the Ideal Parallel to it through a point A.

Let AM (Fig. 98) be the perpendicular to the given line MU, and AU the parallel.

Let the figure be inverted from the point M'', the radius of inversion being the tangent from M'' to the fundamental circle.

Then we obtain a new figure (cf. Fig. 99) in which the corresponding Ideal Lengths are the same, since the circle of inversion is a circle of the system. The lines AM and MU become straight lines through the centre of the fund-

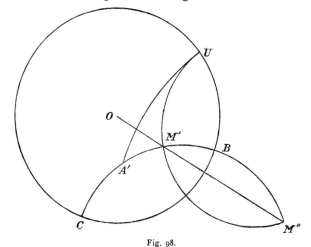

Fig. 98.

amental circle, which is the inverse of the point M'. Also the circle AU becomes the circle $a'u$, touching the radius $m'u$ at u, and cutting $m'a'$ at an angle $\Pi(p)$. These radii, $m'u$, $m'b$, are also Ideal Lines of the system.

The Ideal Length of the Segment AM is taken as p.

Then
$$p = \log\left(\frac{A'B}{A'C}\,\bigg|\,\frac{M'B}{M'C}\right)$$
$$= \log\left(\frac{a'b}{a'c}\,\bigg|\,\frac{m'b}{m'c}\right)$$
$$= \log\left(\frac{a'b}{a'c}\right).$$

But $a'c = k - k\tan\left(\dfrac{\pi}{4} - \dfrac{\Pi(p)}{2}\right)$,

and $a'b = k + k\tan\left(\dfrac{\pi}{4} - \dfrac{\Pi(p)}{2}\right)$, where k is the radius of the fundamental circle.

Thus $p = \log \cot \dfrac{\Pi\,(p)}{2}$;

and $e^{-p} = \tan \dfrac{\Pi\,(p)}{2}$.

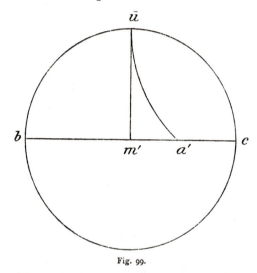

Fig. 99.

Finally, in this geometry there will be three kinds of circles. There will be the *circle*, with its centre at a finite distance; the *Limiting Curve* or *Horocycle*, with its centre at infinity, (at a point where two parallels meet); and the *Equidistant Curve*, with its centre at the *imaginary* point of intersection of two lines with a common perpendicular.

The first of these curves would be traced out in the Ideal Geometry by one end of an Ideal Segment, when it is reflected in the lines passing through the other end; that is, by the rotation of this Ideal Segment about that end. The second occurs when the Ideal Segment is reflected in the successive lines of the pencil of Ideal Lines all parallel to it in the same direction; and the third, when the reflection takes place in the system of Ideal Lines which all have a perpendicular with this segment. That these correspond to the common *Circle*, the *Horocycle* and the *Equidistant Curve* of the Hyperbolic Geometry is easily proved.

§ 12. *The Impossibility of Proving Euclid's Parallel Postulate.*

We could obtain other results of the Hyperbolic Geometry, and find some of its constructions, by further examination of the properties of this set of circles; but this is not our object. Our argument was directed to proving, by reasoning involving only elementary geometry, that it is impossible for any inconsistency or contradiction to arise in this Non-Euclidean Geometry. If such contradiction entered into this Plane Geometry, it would also occur in the interpretation of the result in the Ideal Geometry. Thus the contradiction would also be found in the Euclidean Geometry. We can, therefore, state that it is impossible that any logical inconsistency could be traced in the Plane Hyperbolic Geometry. It could still be argued that such contradiction might be found in the Solid Hyperbolic Geometry. An answer to this objection is at once forthcoming. The geometry of the system of circles, all orthogonal to a fixed circle, can be at once extended into a three dimensional system. The *Ideal Points* are taken as the pairs of points inverse to a fixed sphere, excluding the points on the surface of the sphere from their domain. The *Ideal Lines* are the circles through two Ideal Points. The *Ideal Planes* are the spheres through three Ideal Points, not lying on an Ideal Line. The ordinary plane enters as a particular case of these Ideal Planes, and so the Plane Geometry just discussed is a special case of a plane geometry on this system. With suitable definitions of Ideal Lengths, Ideal Parallels and Ideal Displacements, we have a Solid Geometry exactly analogous to the Hyperbolic Solid

Geometry. It follows that no logical inconsistency can exist in the Hyperbolic Solid Geometry, since if there were such a contradiction, it would also be found in the interpretation of the result in this Ideal Geometry; and therefore it would enter into the Euclidean Geometry.

By this result our argument is complete. However far the Hyperbolic Geometry were developed, no contradictory results could be obtained. This system is thus logically possible; and the axioms upon which it is founded are not contradictory. Hence it is impossible to prove Euclid's Parallel Postulate, since its proof would involve the denial of the Parallel Postulate of BOLYAI-LOBATSCHEWSKY.

Index of Authors.

[The numbers refer to pages.]

THE SCIENCE OF
ABSOLUTE SPACE

BY

JOHN BOLYAI

TRANSLATED BY

DR. GEORGE BRUCE HALSTED

TRANSLATOR'S INTRODUCTION.

The immortal *Elements* of Euclid was already in dim antiquity a classic, regarded as absolutely perfect, valid without restriction.

Elementary geometry was for two thousand years as stationary, as fixed, as peculiarly Greek, as the Parthenon. On this foundation pure science rose in Archimedes, in Apollonius, in Pappus; struggled in Theon, in Hypatia; declined in Proclus; fell into the long decadence of the Dark Ages.

The book that monkish Europe could no longer understand was then taught in Arabic by Saracen and Moor in the Universities of Bagdad and Cordova.

To bring the light, after weary, stupid centuries, to western Christendom, an Englishman, Adelhard of Bath, journeys, to learn Arabic, through Asia Minor, through Egypt, back to Spain. Disguised as a Mohammedan student, he got into Cordova about 1120, obtained a Moorish copy of Euclid's *Elements,* and made a translation from the Arabic into Latin.

The first printed edition of Euclid, published in Venice in 1482, was a Latin version from the Arabic. The translation into Latin from the Greek, made by Zamberti from a MS. of Theon's revision, was first published at Venice in 1505.

Twenty-eight years later appeared the *editio princeps* in Greek, published at Basle in 1533 by John Hervagius, edited by Simon Grynaeus. This was for a century and three-quarters the only printed Greek text of all the books, and from it the first English translation (1570) was made by "Henricus Billingsley," afterward Sir Henry Billingsley, Lord Mayor of London in 1591.

And even to-day, 1895, in the vast system of examinations carried out by the British Government, by Oxford, and by Cambridge, no proof of a theorem in geometry will be accepted which infringes Euclid's sequence of propositions.

Nor is the work unworthy of this extraordinary immortality.

Says Clifford: "This book has been for nearly twenty-two centuries the encouragement and guide of that scientific thought which is one thing with the progress of man from a worse to a better state.

"The encouragement; for it contained a body of knowledge that was really known and could be relied on.

"The guide; for the aim of every student of every subject was to bring his knowledge of that subject into a form as perfect as that which geometry had attained."

But Euclid stated his assumptions with the most painstaking candor, and would have smiled at the suggestion that he claimed for his conclusions any other truth than perfect deduction from assumed hypotheses. In favor of the external reality or truth of those assumptions he said no word.

Among Euclid's assumptions is one differing from the others in prolixity, whose place fluctuates in the manuscripts.

Peyrard, on the authority of the Vatican MS., puts it among the postulates, and it is often called the parallel-postulate. Heiberg, whose edition of the text is the latest and best (Leipzig, 1883-1888), gives it as the fifth postulate.

James Williamson, who published the closest translation of Euclid we have in English, indicating, by the use of italics, the words not in the original, gives this assumption as eleventh among the Common Notions.

Bolyai speaks of it as Euclid's Axiom XI.
Todhunter has it as twelfth of the Axioms.

Clavius (1574) gives it as Axiom 13.

The Harpur Euclid separates it by forty-
eight pages from the other axioms.

It is not used in the first twenty-eight propositions of Euclid. Moreover, when at length
used, it appears as the inverse of a proposition
already demonstrated, the seventeenth, and is
only needed to prove the inverse of another
proposition already demonstrated, the twenty-
seventh.

Now the great Lambert expressly says that
Proklus demanded a proof of this assumption
because when inverted it is demonstrable.

All this suggested, at Europe's renaissance,
not a doubt of the necessary external reality
and exact applicability of the assumption, but
the possibility of deducing it from the other
assumptions and the twenty-eight propositions
already proved by Euclid without it.

Euclid demonstrated things more axiomatic
by far. He proves what every dog knows,
that any two sides of a triangle are together
greater than the third.

Yet after he has finished his demonstration,
that straight lines making with a transversal
equal alternate angles are parallel, in order to

prove the inverse, that parallels cut by a transversal make equal alternate angles, he brings in the unwieldy assumption thus translated by Williamson (Oxford, 1781):

"11. And if a straight line meeting two straight lines make those angles which are inward and upon the same side of it less than two right angles, the two straight lines being produced indefinitely will meet each other on the side where the angles are less than two right angles."

As Staeckel says, "it requires a certain courage to declare such a requirement, alongside the other exceedingly simple assumptions and postulates." But was courage likely to fail the man who, asked by King Ptolemy if there were no shorter road in things geometric than through his *Elements?* answered, "To geometry there is no special way for kings!"

In the brilliant new light given by Bolyai and Lobachevski we now see that Euclid understood the crucial character of the question of parallels.

There are now for us no better proofs of the depth and systematic coherence of Euclid's masterpiece than the very things which, their cause unappreciated, seemed the most noticeable blots on his work.

Sir Henry Savile, in his Praelectiones on Euclid, Oxford, 1621, p. 140, says: "In pulcherrimo Geometriae corpore duo sunt naevi, duae labes . . ." etc., and these two blemishes are the theory of parallels and the doctrine of proportion; the very points in the Elements which now arouse our wondering admiration. But down to our very nineteenth century an ever renewing stream of mathematicians tried to wash away the first of these supposed stains from the most beauteous body of Geometry.

The year 1799 finds two extraordinary young men striving thus

"To gild refined gold, to paint the lily,
To cast a perfume o'er the violet."

At the end of that year Gauss from Braunschweig writes to Bolyai Farkas in Klausenburg (Kolozsvár) as follows: [Abhandlungen der Koeniglichen Gesellschaft der Wissenschaften zu Goettingen, Bd. 22, 1877.]

"I very much regret, that I did not make use of our former proximity, to find out *more* about your investigations in regard to the first grounds of geometry; I should certainly thereby have spared myself much vain labor, and would have become more restful than any one, such as I, can be, so long as on such a subject there yet remains so much to be wished for.

In my own work thereon I myself have advanced far (though my other wholly heterogeneous employments leave me little time therefor) but *the* way, which I have hit upon, leads not so much to the goal, which one wishes, as much more to making doubtful the truth of geometry.

Indeed I have come upon much, which with most no doubt would pass for a proof, but which in my eyes proves as good as *nothing*.

For example, if one could prove, that a rectilineal triangle is possible, whose content may be greater, than any given surface, then I am in condition, to prove with perfect rigor all geometry.

Most would indeed let that pass as an axiom; I not; it might well be possible, that, how far apart soever one took the three vertices of the triangle in space, yet the content was always under a given limit.

I have more such theorems, but in none do I find anything satisfying."

From this letter we clearly see that in 1799 Gauss was still trying to prove that Euclid's is the only non-contradictory system of geometry, and that it is the system regnant in the external space of our physical experience.

The first is false; the second can never be proven.

Before another quarter of a century, Bolyai János, then unborn, had created another possible universe; and, strangely enough, though nothing renders it impossible that the space of our physical experience may, this very year, be satisfactorily shown to belong to Bolyai János, yet the same is not true for Euclid.

To decide our space is Bolyai's, one need only show a single rectilineal triangle whose angle-sum measures less than a straight angle. And this could be shown to exist by imperfect measurements, such as human measurements must always be. For example, if our instruments for angular measurement could be brought to measure an angle to within one millionth of a second, then if the lack were as great as two millionths of a second, we could make certain its existence.

But to prove Euclid's system, we must show that a triangle's angle-sum is *exactly* a straight angle, which nothing human can ever do.

However this is anticipating, for in 1799 it seems that the mind of the elder Bolyai, Bolyai Farkas, was in precisely the same state as

that of his friend Gauss. Both were intensely
trying to prove what now we know is indemonstrable.   And perhaps Bolyai got nearer
than Gauss to the unattainable. In his "Kurzer
Grundriss eines Versuchs," etc., p. 46, we read:
"Koennten jede 3 Punkte, die nicht in einer
Geraden sind, in eine Sphaere fallen, so waere
das Eucl. Ax. XI. bewiesen." Frischauf calls
this "das anschaulichste Axiom."   But in his
Autobiography written in Magyar, of which
my Life of Bolyai contains the first translation ever made, Bolyai Farkas says: "Yet I
could not become satisfied with my different
treatments of the question of parallels, which
was ascribable to the long discontinuance of
my studies, or more probably it was due to
myself that I drove this problem to the point
which robbed my rest, deprived me of tran-
quillity."

It is wellnigh certain that Euclid tried his
own calm, immortal genius, and the genius of
his race for perfection, against this self-same
question. If so, the benign intellectual pride
of the founder of the mathematical school of
the greatest of universities, Alexandria, would
not let the question cloak itself in the obscuri-
ties of the infinitely great or the infinitely
small. He would say to himself: "Can I prove

this plain, straightforward, simple theorem:
"those straights which are produced indefinitely from less than two right angles meet."
[This is the form which occurs in the Greek
of Eu. I. 29.]

Let us not underestimate the subtle power
of that old Greek mind. We can produce no
Venus of Milo. Euclid's own treatment of
proportion is found as flawless in the chapter
which Stolz devotes to it in 1885 as when
through Newton it first gave us our present
continuous number-system.

But what fortune had this genius in the fight
with its self-chosen simple theorem? Was it
found to be deducible from all the definitions,
and the nine "Common Notions," and the five
other Postulates of the immortal Elements?
Not so. But meantime Euclid went ahead
without it through twenty-eight propositions,
more than half his first book. But at last
came the practical pinch, then as now the triangle's angle-sum.

He gets it by his twenty-ninth theorem: "A
straight falling upon two parallel straights
makes the alternate angles equal."

But for the proof of this he needs that re-
calcitrant proposition which has how long
been keeping him awake nights and waking

him up mornings?   Now at last, true man of
science, he acknowledges it indemonstrable by
spreading it in all its ugly length among his
postulates.

Since Schiaparelli has restored the astronomical system of Eudoxus, and Hultsch has
published the writings of Autolycus, we see
that Euclid knew surface-spherics, was familiar with triangles whose angle-sum is more
than a straight angle.   Did he ever think to
carry out for himself the beautiful system of
geometry which comes from the contradiction
of his indemonstrable postulate; which exists
if there be straights produced indefinitely from
less than two right angles yet nowhere meeting; which is real if the triangle's angle-sum
is less than a straight angle?

Of how naturally the three systems of geometry flow from just exactly the attempt we
suppose Euclid to have made, the attempt to
demonstrate his postulate fifth, we have a most
romantic example in the work of the Italian
priest, Saccheri, who died the twenty-fifth of
October, 1733.   He studied Euclid in the edition of Clavius, where the fifth postulate is
given as Axiom 13.   Saccheri says it should
not be called an axiom, but ought to be demonstrated.   He tries this seemingly simple

task; but his work swells to a quarto book of 101 pages.

Had he not been overawed by a conviction of the absolute necessity of Euclid's system, he might have anticipated Bolyai János, who ninety years later not only discovered the new world of mathematics but appreciated the transcendent import of his discovery.

Hitherto what was known of the Bolyais came wholly from the published works of the father Bolyai Farkas, and from a brief article by Architect Fr. Schmidt of Budapest "Aus dem Leben zweier ungarischer Mathematiker, Johann und Wolfgang Bolyai von Bolya." Grunert's Archiv, Bd. 48, 1868, p. 217.

In two communications sent me in September and October 1895, Herr Schmidt has very kindly and graciously put at my disposal the results of his subsequent researches, which I will here reproduce. But meantime I have from entirely another source come most unexpectedly into possession of original documents so extensive, so precious that I have determined to issue them in a separate volume devoted wholly to the life of the Bolyais; but these are not used in the sketch here given.

Bolyai Farkas was born February 9th, 1775, at Bolya, in that part of Transylvania (Er-

dély) called Székelyföld. He studied first at
Enyed, afterward at Klausenburg (Kolozsvár),
then went with Baron Simon Kemény to Jena
and afterward to Goettingen. Here he met
Gauss, then in his 19th year, and the two
formed a friendship which lasted for life.

The letters of Gauss to his friend were sent
by Bolyai in 1855 to Professor Sartorius von
Walterhausen, then working on his biography
of Gauss. Everyone who met Bolyai felt that
he was a profound thinker and a beautiful
character.

Benzenberg said in a letter written in 1801
that Bolyai was one of the most extraordinary
men he had ever known.

He returned home in 1802, and in January,
1804, was made professor of mathematics in
the Reformed College of Maros-Vásárhely.
Here for 47 years of active teaching he had
for scholars nearly all the professors and no-
bility of the next generation in Erdély.

Sylvester has said that mathematics is poesy.

Bolyai's first published works were dramas.

His first published book on mathematics was
an arithmetic:

Az arithmetica eleje. 8vo. i–xvi, 1–162 pp.
The copy in the library of the Reformed College is enriched with notes by Bolyai János.

Next followed his chief work, to which he constantly refers in his later writings. It is in Latin, two volumes, 8vo, with title as follows:

TENTAMEN | JUVENTUTEM STUDIOSAM | IN ELEMENTA MATHESĘOS PURAE, ELEMENTARIS AC | SUBLIMIORIS, METHODO INTUITIVA, EVIDENTIA— | QUE HUIC PROPRIA, INTRODUCENDI. |

CUM APPENDICE TRIPLICI. | Auctore Professore Matheseos et Physices Chemiaeque | Publ. Ordinario. | Tomus Primus. | *Maros Vasarhelyini.* 1832. | Typis Collegii Reformatorum per JOSEPHUM, et | SIMEONEM KALI de felsö Vist. | At the back of the title: Imprimatur. | M. Vásárhelyini Die | 12 Octobris, 1829. | Paulus Horváth m. p. | Abbas, Parochus et Censor | Librorum.

Tomus Secundus. | *Maros Vasarhelyini.* 1833. |

The first volume contains:

Preface of two pages: *Lectori salutem.*

A folio table: *Explicatio signorum.*

Index rerum (I—XXXII). *Errata* (XXXIII—XXXVII).

Pro tyronibus prima vice legentibus notanda sequentia (XXXVIII—LII).

Errores (LIII—LXVI).

Scholion (LXVII—LXXIV).

Plurium errorum haud animadversorum numerus minuitur (LXXV—LXXVI).

Recensio per auctorem ipsum facta (LXXVII—LXXVIII).

Errores recentius detecti (L X X V — XCVIII).

Now comes the body of the text (pages 1—502).

Then, with special paging, and a new title page, comes the immortal Appendix, here given in English.

Professors Staeckel and Engel make a mistake in their "Parallellinien" in supposing that this Appendix is referred to in the title of "Tentamen." On page 241 they quote this title, including the words "Cum appendice triplici," and say: "In dem dritten Anhange, der nur 28 Seiten umfasst, hat Johann Bolyai seine neue Geometrie entwickelt."

It is not a third Appendix, nor is it referred to at all in the words "Cum appendice triplici."

These words, as explained in a prospectus in the Magyar language, issued by Bolyai Farkas, asking for subscribers, referred to a real triple Appendix, which appears, as it should, at the end of the book Tomus Secundus, pp. 265–322.

The now world renowned Appendix by Bolyai János was an afterthought of the father, who prompted the son not "to occupy himself with the theory of parallels," as Staeckel says, but to translate from the German into Latin a condensation of his treatise, of which the principles were discovered and properly appreciated in 1823, and which was given in writing to Johann Walter von Eckwehr in 1825.

The father, without waiting for Vol. II, inserted this Latin translation, with separate paging (1–26), as an Appendix to his Vol. I, where, counting a page for the title and a page "Explicatio signorum," it has twenty-six numbered pages, followed by two unnumbered pages of Errata.

The treatise itself, therefore, contains only twenty-four pages — the most extraordinary two dozen pages in the whole history of thought!

Milton received but a paltry £5 for his Paradise Lost; but it was at least plus £5.

Bolyai János, as we learn from Vol. II, p. 384, of "*Tentamen*," contributed for the printing of his eternal twenty-six pages, 104 florins 50 kreuzers.

That this Appendix was finished considerably before the Vol. I, which it follows, is seen from the references in the text, breathing a just admiration for the Appendix and the genius of its author.

Thus the father says, p. 452: Elegans est conceptus *similium,* quem J. B. *Appendicis Auctor* dedit. Again, p. 489: *Appendicis Auctor,* rem acumine singulari aggressus, Geometriam pro omni casu absolute veram posuit; quamvis e magna mole, tantum summe necessaria, in Appendice hujus tomi exhibuerit, multis (ut tetraedri resolutione generali, pluribusque aliis disquisitionibus elegantibus) brevitatis studio omissis.

And the volume ends as follows, p. 502: Nec operae pretium est plura referre; quum res tota exaltiori contemplationis puncto, in ima penetranti oculo, tractetur in Appendice sequente, a quovis fideli veritatis purae alumno diagna legi.

The father gives a brief resumé of the results of his own determined, life-long, desperate efforts to do that at which Saccheri, J. H. Lambert, Gauss also had failed, to establish Euclid's theory of parallels *a priori.*

He says, p. 490: "'Tentamina idcirco quae olim feceram, breviter exponenda veniunt; ne saltem alius quis operam eandem perdat." He anticipates J. Delboeuf's "Prolégoménes philosophiques de la géométrie et solution des postulats," with the full consciousness in addition that it is *not* the solution,—that the final solution has crowned not his own intense efforts, but the genius of his son.

This son's Appendix which makes all preceding space only a special case, only a species under a genus, and so requiring a descriptive adjective, *Euclidean*, this wonderful production of pure genius, this strange Hungarian flower, was saved for the world after more than thirty-five years of oblivion, by the rare erudition of Professor Richard Baltzer of Dresden, afterward professor in the University of Giessen. He it was who first did justice publicly to the works of Lobachevski and Bolyai.

Incited by Baltzer, in 1866 J. Hoüel issued a French translation of Lobachevski's Theory of Parallels, and in a note to his Preface says: "M. Richard Baltzer, dans la seconde édition de ses excellents *Elements de Geometrie,* a, le premier, introduit ces notions exactes à la place qu'elles doivent occuper." Honor to

Baltzer! But alas! father and son were already in their graves!

Fr. Schmidt in the article cited (1868) says: "It was nearly forty years before these profound views were rescued from oblivion, and Dr. R. Baltzer, of Dresden, has acquired imperishable titles to the gratitude of all friends of science as the first to draw attention to the works of Bolyai, in the second edition of his excellent Elemente der Mathematik (1866–67). Following the steps of Baltzer, Professor Hoüel, of Bordeaux, in a brochure entitled, Essai critique sur les principes fondamentaux de la Géométrie élémentaire, has given extracts from Bolyai's book, which will help in securing for these new ideas the justice they merit."

The father refers to the son's Appendix again in a subsequent book, Urtan elemei kezdöknek [Elements of the science of space for beginners] (1850–51 , pp. 48. In the College are preserved three sets of figures for this book, two by the author and one by his grandson, a son of János.

The last work of Bolyai Farkas, the only one composed in German, is entitled,

Kurzer Grundriss eines Versuchs

I. Die Arithmetik, durch zvekmässig kons-

truirte Begriffe, von eingebildeten und unend-
lich-kleinen Grössen gereinigt, anschaulich
und logisch-streng darzustellen.

II. In der Geometrie, die Begriffe der geraden Linie, der Ebene, des Winkels allgemein,
der winkellosen Formen, und der Krummen,
der verschiedenen Arten der Gleichheit u. d.
gl. nicht nur scharf zu bestimmen; sondern
auch ihr Seyn im Raume zu beweisen: und da
die Frage, *ob zwey von der dritten geschnittene Geraden, wenn die summe der inneren
Winkel nicht = 2R, sich schneiden oder
nicht?* neimand auf der Erde ohne ein Axiom
(wie Euklid das XI) aufzustellen, beantworten
wird; die davon unabhängige Geometrie abzusondern; und eine auf die *Ja*—Antwort,
andere auf das *Nein* so zu bauen, das die
Formeln der letzten, auf ein Wink auch in der
ersten gültig seyen.

Nach ein lateinischen Werke von 1829, M.
Vásárhely, und eben daselbst gedruckten un-
grischen.

Maros Vásárhely 1851. 8vo. pp. 88.

In this book he says, referring to his son's
Appendix: "Some copies of the work published here were sent at that time to Vienna,
to Berlin, to Goettingen. . . . From Goettingen the giant of mathematics, who from

his pinnacle embraces in the same view the stars and the abysses, wrote that he was surprised to see accomplished what he had begun, only to leave it behind in his papers.''

This refers to 1832. The only other record that Gauss ever mentioned the book is a letter from Gerling, written October 31st, 1851, to Wolfgang Boylai, on receipt of a copy of ''Kurzer Grundriss.'' Gerling, a scholar of Gauss, had been from 1817 Professor of Astronomy at Marburg. He writes: ''I do not mention my earlier occupation with the theory of parallels, for already in the year 1810–1812 with Gauss, as earlier 1809 with J. F. Pfaff I had learned to perceive how all previous attempts to prove the Euclidean axiom had miscarried. I had then also obtained preliminary knowledge of your works, and so, when I first [1820] had to print something of my view thereon, I wrote it exactly as it yet stands to read on page 187 of the latest edition.

''We had about this time [1819] here a law professor, Schweikart, who was formerly in Charkov, and had attained to similar ideas, since without help of the Euclidean axiom he developed in its beginnings a geometry which he called Astralgeometry. What he communicated to me thereon I sent to Gauss, who then informed me how much farther already had been attained on this way, and later also expressed himself about the great acquisition, which is offered to the few expert judges in the Appendix to your book."

The "latest edition" mentioned appeared in 1851, and the passage referred to is: "This proof [of the parallel-axiom] has been sought in manifold ways by acute mathematicians, but yet until now not found with complete sufficiency. So long as it fails, the theorem, as all founded on it, remains a hypothesis, whose validity for our life indeed is sufficiently proven by *experience,* whose *general, necessary exactness,* however, could be doubted without absurdity."

Alas! that this feeble utterance should have seemed sufficient for more than thirty years to the associate of Gauss and Schweikart, the latter certainly one of the independent discoverers of the non-Euclidean geometry. But then, since neither of these sufficiently realized the transcendent importance of the matter to publish any of their thoughts on the subject, a more adequate conception of the issues at stake could scarcely be expected of the scholar and colleague. How different with Bolyai János and Lobachévski, who claimed at once, unflinchingly, that their discovery marked an epoch in human thought so momentous as to be unsurpassed by anything recorded in the history of philosophy or of science, demonstrating as had never been proven before the supremacy of pure reason at the very moment of overthrowing what had forever seemed its surest possession, the axioms of geometry.

On the 9th of March, 1832, Bolyai Farkas was made corresponding member in the mathematics section of the Magyar Academy.

As professor he exercised a powerful influence in his country.

In his private life he was a type of true originality. He wore roomy black Hungarian pants, a white flannel jacket, high boots, and a broad hat like an old-time planter's. The smoke-stained wall of his antique domicile was adorned by pictures of his friend Gauss, of Schiller, and of Shakespeare, whom he loved to call the child of nature. His violin was his constant solace.

He died November 20th, 1856. It was his wish that his grave should bear no mark.

The mother of Bolyai János, née Arkosi Benkö Zsuzsanna, was beautiful, fascinating, of extraordinary mental capacity, but always nervous.

János, a lively, spirited boy, was taught mathematics by his father. His progress was marvelous. He required no explanation of theorems propounded, and made his own demonstrations for them, always wishing his father to go on. "Like a demon, he always pushed me on to tell him more."

At 12, having passed the six classes of the Latin school, he entered the philosophic-curriculum, which he passed in two years with great distinction.

When about 13, his father, prevented from meeting his classes, sent his son in his stead. The students said they liked the lectures of the son better than those of the father. He already played exceedingly well on the violin.

In his fifteenth year he went to Vienna to K. K. Ingenieur-Akademie.

In August, 1823, he was appointed "sous-lieutenant" and sent to Temesvár, where he was to present himself on the 2nd of September.

From Temesvár, on November 3rd, 1823, János wrote to his father a letter in Magyar, of which a French translation was sent me by Professor Koncz József on February 14th,

1895. This will be given in full in my life of Bolyai; but here an extract will suffice:

"My Dear and Good Father:
"I have so much to write about my new inventions that it is impossible for the moment to enter into great details, so I write you only on one-fourth of a sheet. I await your answer to my letter of two sheets; and perhaps I would not have written you before receiving it, if I had not wished to address to you the letter I am writing to the Baroness, which letter I pray you to send her.

"First of all I reply to you in regard to the binominal.

* * * * * * * * *

"Now to something else, so far as space permits. I intend to write, as soon as I have put it into order, and when possible to publish, a work on parallels.

"At this moment it is not yet finished, but the way which I have followed promises me with certainty the attainment of the goal, if it in general is attainable. It is not yet attained, but I have discovered such magnificent things that I am myself astonished at them.

"It would be damage eternal if they were

lost. When you see them, my father, you yourself will acknowledge it. Now I can not say more, only so much: *that from nothing I have created another wholly new world.* All that I have hitherto sent you compares to this only as a house of cards to a castle.

"P. S.—I dare to judge absolutely and with conviction of these works of my spirit before you, my father; I do not fear from you any false interpretation (that certainly I would not merit), which signifies that, in certain regards, I consider you as a second self."

From the Bolyai MSS., now the property of the College at Maros-Vásárhely, Fr. Schmidt has extracted the following statement by János:

"First in the year 1823 have I pierced through the problem in its essence, though also afterwards completions yet were added.

"I communicated in the year 1825 to my former teacher, Herr Johann Walter von Eckwehr (later k. k. General) [in the Austrian Army], a written treatise, which is still in his hands.

"On the prompting of my father I translated my treatise into the Latin language, and

it appeared as *Appendix* to the *Tentamen,*
1832."

The profound mathematical ability of Bolyai János showed itself physically not only in
his handling of the violin, where he was a
master, but also of arms, where he was unap-
proachable.

It was this skill, combined with his haughty
temper, which caused his being retired as Cap-
tain on June 16th, 1833, though it saved him
from the fate of a kindred spirit, the lamented
Galois, killed in a duel when only 19.   Bolyai,
when in garrison with cavalry officers, was
provoked by thirteen of them and accepted all
their challenges on condition that he be permitted after each duel to play a bit on his
violin.   He came out victor from his thirteen
duels, leaving his thirteen adversaries on the
square.

He projected a universal language for
speech as we have it for music and for mathematics.

He left parts of a book entitled: Principia
doctrinae novae quantitatum imaginariarum
perfectae uniceque satisfacientis, aliaeque dis-
quisitiones analyticae et analytico - geome-
tricae cardinales gravissimaeque; auctore

Johan. Bolyai de eadem, C. R. austriaco castrensium captaneo pensionato.

Vindobonae vel Maros Vásárhelyini, 1853.

Bolyai Farkas was a student at Goettingen
from 1796 to 1799.

In 1799 he returned to Kolozsvár, where
Bolyai János was born December 18th, 1802.

He died January 27th, 1860, four years
after his father.

In 1894 a monumental stone was erected on
his long-neglected grave in Maros-Vásárhely
by the Hungarian Mathematico-Physical Society.

APPENDIX.

SCIENTIAM SPATII *absolute veram* exhibens:

a veritate aut falsitate Axiomatis XI *Euclidei*
(a priori haud unquam decidenda) in-
dependentem: adjecta ad casum falsitatis, quadratura circuli
geometrica.

———•———

Auctore JOHANNE BOLYAI de eadem, Geometrarum
in Exercitu Caesareo Regio Austriaco
Castrensium Capitaneo.

EXPLANATION OF SIGNS.

The straight AB means the aggregate of all points situated
in the same straight line with A and B.

The sect AB means that piece of the straight AB between
the points A and B.

The ray AB means that half of the straight AB which com-
mences at the point A and contains the point B.

The plane ABC means the aggregate of all points situated
in the same plane as the three points (not in a
straight) A, B, C.

The hemi-plane ABC means that half of the plane ABC
which starts from the straight AB and contains the
point C.

ABC means the smaller of the pieces into which the plane
ABC is parted by the rays BA, BC, or the non-reflex
angle of which the sides are the rays BA, BC.

ABCD (the point D being situated within $\angle$ ABC, and the
straights BA, CD not intersecting) means the portion
of $\angle$ ABC comprised between ray BA, sect BC, ray
CD; while BACD designates the portion of the plane
ABC comprised between the straights AB and CD.

$\perp$ is the sign of perpendicularity.

$\parallel$ is the sign of parallelism.

$\angle$ means angle.

rt. $\angle$ is right angle.

st. $\angle$ is straight angle.

$\cong$ is the sign of congruence, indicating that two magnitudes are superposable.

AB $\doteqdot$ CD means $\angle$ CAB $= \angle$ ACD.

$x \doteq a$ means x converges toward the limit a.

$\triangle$ is triangle.

$\odot r$ means the [circumference of the] circle of radius r.

area $\odot r$ means the area of the surface of the circle of radius r.

THE SCIENCE ABSOLUTE OF SPACE.

§1. If the ray AM is not cut by the ray [3] BN, situated in the same plane, but is cut by every ray BP comprised in the angle ABN, we will call ray BN *parallel* to ray AM; this is designated by BN ‖ AM.

It is evident that *there is one such ray BN, and only one,* passing through any point B (taken outside of the straight AM), and that the sum of the angles BAM, ABN can not exceed a st. ∠; for in moving BC around B until BAM+ABC=st. ∠, somewhere ray BC *first* does not cut ray AM, and it is then BC ‖ AM. It is clear that BN ‖ EM, wherever the point E be taken on the straight AM (supposing in all such cases AM>AE).

If while the point C goes away to infinity on ray AM, always CD=CB, we will have constantly CDB=(CBD<NBC); but NBC≐0; and so also ADB≐0.

§2.   If BN ‖ AM, we will have also CN ‖ AM.

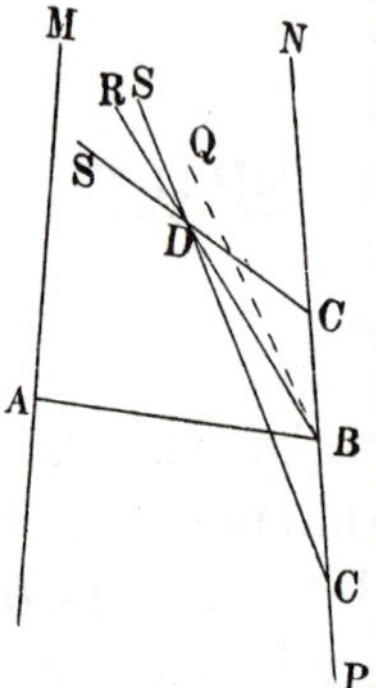

For take D anywhere in MACN. If C is on ray BN, ray BD cuts ray AM, since BN ‖ AM, and so also ray CD cuts ray AM. But if C is on ray BP, take BQ ‖ CD; BQ falls within the ∠ ABN (§1), and cuts ray AM; and so also ray CD cuts ray AM. Therefore every ray CD (in ACN) cuts, in each case, the ray AM, without CN itself cutting ray AM. Therefore always CN ‖ AM.

FIG. 2.

§3.   (Fig. 2.) If BR and CS and each ‖ AM, and C is not on the ray BR, then ray BR and ray CS do not intersect. For if ray BR and ray CS had a common point D, then (§ 2) DR and DS would be each ‖ AM, and ray DS (§ 1) would fall on ray DR, and C on the ray BR (contrary to the hypothesis).

§ 4.   If MAN>MAB, we will have for every point B of ray AB, a point C of ray AM, such that BCM=NAM.

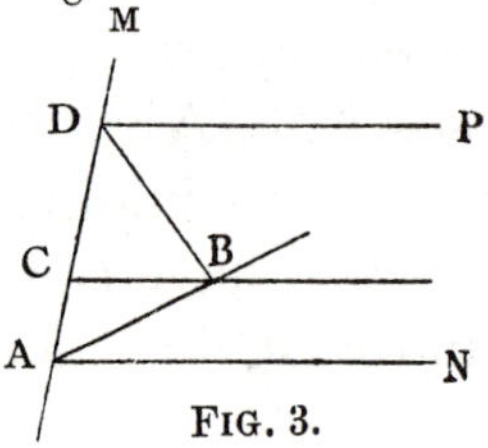

For (by § 1) is granted BDM>NAM, and so that MDP=MAN, and B falls in

FIG. 3.

NADP. If therefore NAM is carried along AM until ray AN arrives on ray DP, ray AN will somewhere have necessarily passed through B, and some BCM=NAM.

§ 5. If BN ‖ AM, there is on the straight [4] AM a point F such that FM≏BN.

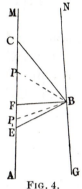

FIG. 4.

For by §1 is granted BCM>CBN; and if CE=CB, and so EC≏BC; evidently BEM<EBN. The point P is moved on EC, the angle BPM always being called u, and the angle PBN always v; evidently u is at first less than the corresponding v, but afterwards greater. Indeed u increases *continuously* from BEM to BCM; since (by §4) there exists *no* angle >BEM and <BCM, to which u does not at some time become equal. Likewise v decreases continuously from EBN to CBN. There is therefore on EC a point F such that BFM=FBN.

§6. If BN ‖ AM and E anywhere in the straight AM, and G in the straight BN; then GN ‖ EM and EM ‖ GN. For (by §1) BN ‖ EM, whence (by §2) GN ‖ EM. If moreover FM≏ BN (§5); then MFBN≅NBFM, and consequently (since BN ‖ FM) also FM ‖ BN, and (by what precedes) EM ‖ GN.

§ 7. If BN and CP are each ‖ AM, and C not on the straight BN; also BN ‖ CP. For the rays BN and CP do not intersect (§3); but AM, BN and CP either are or are not in the same plane; and in the first case, AM either is or is not within BNCP.

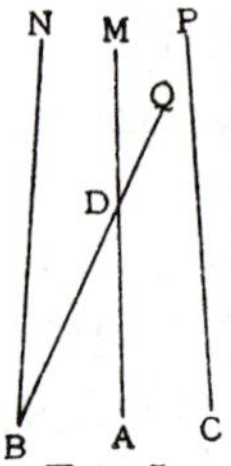

FIG. 5.

If AM, BN, CP are complanar, and AM falls within BNCP; then every ray BQ (in NBC) cuts the ray AM in some point D (since BN ‖ AM); moreover, since DM ‖ CP (§ 6), the ray DQ will cut the ray CP, and so BN ‖ CP.

But if BN and CP are on the same side of AM; then one of them, for example CP, falls between the two other straights BN, AM: but every ray BQ (in NBA) cuts the ray AM, and so also the straight CP. Therefore BN ‖ CP.

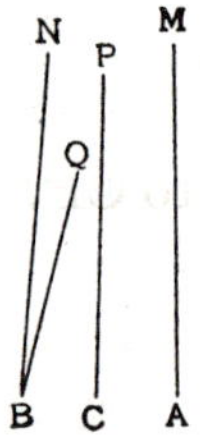

FIG. 6.

If the planes MAB, MAC make *an angle;* then CBN and ABN have in common nothing but the ray BN, while the ray AM (in ABN) and the ray BN, and so also NBC and the ray AM have nothing in common.

But hemi-plane BCD, drawn through any ray BD (in NBA), cuts the ray AM, since ray

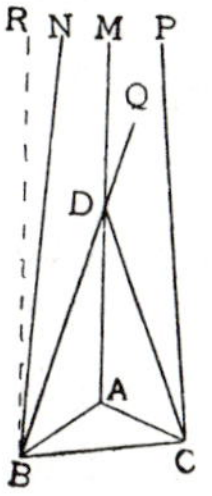

BQ cuts ray AM (as BN ‖ AM).
Therefore in revolving the hemi-plane [5]
BCD around BC until it *begins* to
leave the ray AM, the hemi-plane
BCD at last will fall upon the hemi-
plane BCN. For the same reason this
same will fall upon hemi-plane BCP.

FIG. 7. Therefore BN falls in BCP. Moreover, if BR ‖ CP; then (because also AM ‖ CP)
by like reasoning, BR falls in BAM, and also
(since BR ‖ CP) in BCP. Therefore the
straight BR, being common to the two planes
MAB, PCB, of course is the straight BN, and
hence BN ‖ CP.*

If therefore CP ‖ AM, and B exterior to the
plane CAM; then the intersection BN of the
planes BAM, BCP is ‖ as well to AM as to CP.

§ 8. If BN ‖ and ≏ CP (or more briefly BN
‖ ≏ CP), and AM (in NBCP) bisects
⊥ the sect BC; then BN ‖ AM.

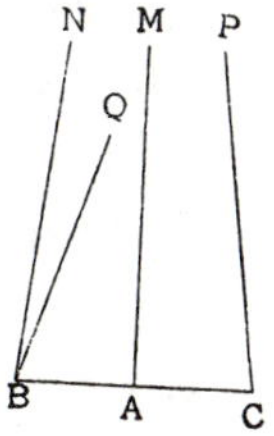

For if ray BN cut ray AM, also
ray CP would cut ray AM at the
same point (because MABN≅
MACP), and this would be common
to the rays BN, CP themselves, al-

FIG. 8.

---

* The third case being put before the other two, these can be
demonstrated together with more brevity and elegance, like case
2 of § 10. [Author's note.]

though BN ‖ CP.   But every ray BQ (in CBN)
cuts ray CP; and so ray BQ cuts also ray AM.
Consequently BN ‖ AN.

§ 9.   If BN ‖ AM, and MAP ⊥ MAB, and the
∠, which NBD makes with
NBA (on that side of MABN,
where MAP is) is <rt.∠; then
MAP and NBD intersect.

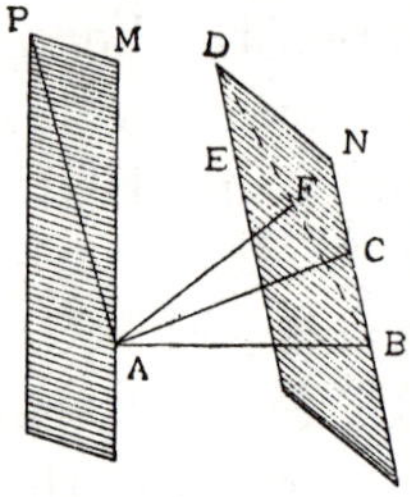

For let ∠BAM=rt.∠, and
AC ⊥ BN (whether or not C
falls on B), and CE ⊥ BN (in
NBD); by hypothesis ∠ACE
<rt.∠, and AF (⊥ CE) will fall in ACE.

Fig. 9.

Let ray AP be the intersection of the hemi-
planes ABF, AMP (which have the point A
common); since BAM ⊥ MAP, ∠BAP=∠BAM
=rt.∠.

If finally the hemi-plane ABF is placed upon
the hemi-plane ABM (A and B remaining), ray
AP will fall on ray AM; and since AC ⊥ BN,
and sect AF<sect AC, evidently sect AF will
terminate within ray BN, and so BF falls in
ABN.   But in *this* position, ray BF cuts ray AP
(because BN ‖ AM); and so ray AP and ray BF
[6] intersect also in *the original* position; and the
point of section is common to the hemi-planes
MAP and NBD.   Therefore the hemi-planes
MAP and NBD intersect.   Hence follows eas-

ily that the hemi-planes MAP and NBD intersect if the sum of the interior angles which
they make with MABN is < st. ∠.

§ 10. If both BN and CP ‖ ≏ AM; also is

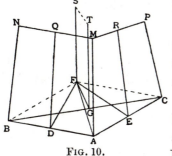

FIG. 10.

BN ‖ ≏ CP.

For either MAB
and MAC make an
angle, or they are in
a plane.

If the first; let the
hemi-plane QDF bi-
sect ⊥ sect AB; then
DQ ⊥ AB, and so DQ
‖ AM (§ 8); likewise if hemi-plane ERS bisects
⊥ sect AC, is ER ‖ AM; whence (§ 7) DQ ‖ ER.

Hence follows easily (by § 9), the hemi-
planes QDF and ERS intersect, and have (§ 7)
their intersection FS ‖ DQ, and (on account of
BN ‖ DQ) also FS ‖ BN. Moreover (for any
point of FS) FB=FA=FC, and the straight
FS falls in the plane TGF, bisecting ⊥ sect BC.
But (by § 7) (since FS ‖ BN) also GT ‖ BN.
In the same way is proved GT ‖ CP. Mean-
while GT bisects ⊥ sect BC; and so TGBN≅
TGCP (§ 1), and BN ‖ ≏ CP.

If BN, AM and CP are in a plane, let (falling without this plane) FS ‖ ≏ AM; then (from

what precedes) FS ∥ ≏ both to BN and to CP, and so also BN ∥ ≏ CP.

§ 11. Consider the aggregate of the point A, and *all* points of which any one B is such, that if BN ∥ AM, also BN ≏ AM; call it F; but the intersection of F with any plane containing the sect AM call L.

F has a point, and one only, on any straight ∥ AM; and evidently L is divided by ray AM into two congruent parts.

Call the ray AM *the axis* of L. Evidently also, in any plane containing the sect AM, there is for the *axis* ray AM a single L. Call any L of this sort the L of this ray AM (in the plane considered, being understood). Evidently by revolving L around AM we describe the F of which ray AM is called the axis, and in turn F *may be ascribed to the axis ray AM.*

[7]   § 12. If B is anywhere on the L of ray AM, and BN ∥ ≏ AM (§ 11); then the L of ray AM and the L of ray BN *coincide.* For suppose, in distinction, L′ the L of ray BN. Let C be anywhere in L′, and CP ∥ ≏ BN (§ 11). Since BN ∥ ≏ AM, so CP ∥ ≏ AM (§ 10), and so C also will fall on L. And if C is anywhere on L, and CP ∥ ≏ AM; then CP ∥ ≏ BN (§ 10); and C also falls on L′ (§ 11). Thus L and L′ are the

same; and every ray BN is also axis of L, and between all axes of this L, is ≏.

The same is evident in the same way of F.

§13. If BN ‖ AM, and CP ‖ DQ, and∠BAM +∠ABN=st.∠; then also ∠DCP+∠CDQ= st.∠.

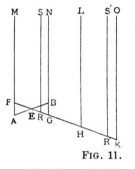

Fig. 11.

For let EA= EB, and EFM= DCP (§4). Since ∠BAM+∠ABN =st. ∠=∠ABN+ ∠ABG, we have ∠EBG=∠EAF; and so if also BG =AF, then△EBG ≅△EAF, ∠BEG=∠AEF and G will fall on the ray FE. Moreover ∠GFM+∠FGN=st.∠ (since ∠EGB=∠EFA).

Also GN ‖ FM (§6).

Therefore if MFRS≅PCDQ, then RS ‖ GN (§7), and R falls within or without the sect FG (unless sect CD=sect FG, where the thing now is evident).

I. In the first case ∠FRS is not >(st.∠−∠ RFM=∠FGN), since RS ‖ FM. But as RS ‖ GN, also ∠FRS is not <∠FGN; and so ∠FRS =∠FGN, and ∠RFM+∠FRS=∠GFM+∠

FGN=st.∠. Therefore also ∠DCP+∠CDQ
=st.∠.

II. If R falls without the sect FG; then
∠NGR=∠MFR, and let MFGN≅NGHL≅
LHKO, and so on, until FK=FR or begins to
be >FR. Then KO ‖ HL ‖ FM (§7).

If K falls on R, then KO falls on RS (§1);
and so ∠RFM+∠FRS=∠KFM+∠FKO=∠
KFM+∠FGN=st.∠; but if R falls within the
sect HK, then (by I) ∠RHL+∠KRS=st.∠=
∠RFM+∠FRS=∠DCP+∠CDQ.

§14. If BN ‖ AM, and CP ‖ DQ, and ∠BAM
+∠ABN<st.∠; then also ∠DCP+∠CDQ<
st.∠.

For if ∠DCP+∠CDQ were not <st.∠, and
so (by §1) were =st.∠, then (by §13) also ∠
BAM+∠ABN=st.∠ (contra hyp.).

§15. Weighing §§13 and 14, *the System of
Geometry resting on the hypothesis of the
truth of Euclid's Axiom XI is called Σ; and
the system founded on the contrary hypoth-*
[8] *esis is S.*

*All things which are not expressly said to
be in Σ or in S, it is understood are enunciated absolutely, that is are asserted true
whether Σ or S is reality.*

§ 16. If AM is the axis of any L; then L, in Σ is a straight ⊥ AM.

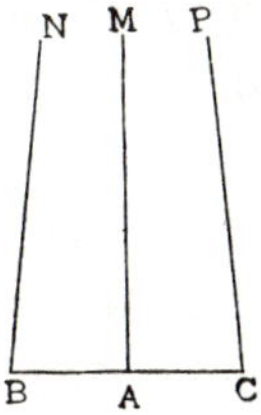

For suppose BN an axis from any point B of L; in Σ, ∠BAM+∠ABN =st.∠, and so ∠BAM=rt.∠.

And if C is any point of the straight AB, and CP ∥ AM; then (by § 13) CP≏AM, and so C on L (§ 11).

Fig. 12.

But in S, no three points A, B, C on L or on F are in a straight. For some one of the axes AM, BN, CP (e. g. AM) falls between the two others; and then (by § 14) ∠BAM and ∠CAM are each <rt.∠.

§ 17. *L in S also is a line, and F a surface.* For (by § 11) any plane ⊥ to the axis ray AM (through any point of F) cuts F in [the circumference of] a circle, of which the plane (by § 14) is ⊥ to no other axis ray BN. If we revolve F about BN, any point of F (by § 12) will remain on F, and the section of F with a plane not ⊥ ray BN will describe a surface; and whatever be the points A, B taken on it, F can so be congruent to itself that A falls upon B (by § 12); therefore F is *a uniform surface.*

Hence evidently (by §§ 11 and 12) L is a uniform line.*

§ **18.**   *The intersection* with F of *any plane*, drawn through a point A of F obliquely to the axis AM, is, in S, *a circle*.

For take A, B, C, three points of this section, and BN, CP, axes; AMBN and AMCP make an angle, for otherwise the plane determined by A, B, C (from § 16) would contain AM, (contra hyp.).   Therefore the planes bisecting ⊥ the sects AB, AC intersect (§ 10) in some *axis* ray FS (of F), and FB=FA=FC.

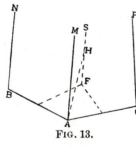

FIG. 13.

Make AH ⊥ FS, and revolve FAH about FS; A will describe a circle of r a d i u s   HA, passing through B and C, and situated *both* in F and in [9] the plane ABC; nor have F and the plane ABC anything in common but ⊙ HA (§ 16).

It is also evident that in revolving the portion FA of the line L (as radius) in F around F, its extremity will describe ⊙ HA.

---

* It is not necessary to restrict the demonstration to the system S; since it may easily be so set forth, that it holds absolutely for S and for Σ.

§ 19. *The perpendicular* BT to the axis BN of L (falling in the plane of L) is, in S,

N

tangent to L. For L has in ray BT no point except B (§ 14), but if BQ falls in TBN, then the center of the section of the plane through BQ perpendicular to TBN with the F of ray BN (§ 18) is evidently located on ray

B

Fig. 14.

BQ; and if sect BQ is a diameter, evidently ray BQ cuts in Q the line L of ray BN.

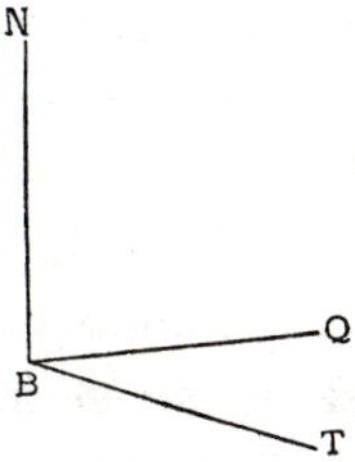

§ 20. Any two points of F determine a line L (§§ 11 and 18); and since (from §§ 16 and 19) L is ⊥ to all its axes, every ∠ of lines L in F is equal to the ∠ of the planes drawn through its sides perpendicular to F.

§ 21. Two L form lines, ray AP and ray BD, in the same F, making with a third L form AB, a sum of interior angles <st.∠, intersect.

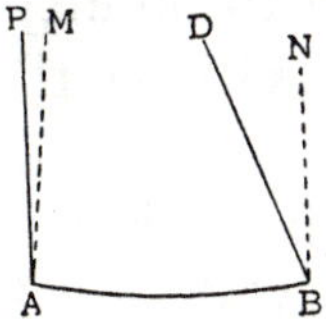

(By line AP in F, is to be understood the line L drawn

Fig. 15.

through A and P, but by ray AP that half of this line beginning at A, in which P falls.)

For if AM, BN are axes of F, then the hemi-planes AMP, BND intersect (§ 9); and F cuts their intersection (by §§ 7 and 11); and so also ray AP and ray BD intersect.

From this it is evident that Euclid's Axiom *XI* and all things which are claimed in geometry and plane trigonometry hold good *absolutely* in F, L lines being substituted in place of straights: therefore the trigonometric functions are taken here in the same sense as in Σ; and the circle of which the L form radius $= r$ in F, is $= 2^{\pi}r;$ and likewise area of $\odot r$ (in F) $= \pi r^2$ (by π understanding $\frac{1}{2}\odot 1$ in F, or the known 3.1415926...)

§ 22. If ray AB were the L of ray AM, and C on ray AM; and the ∠CAB (formed by the straight ray AM and the L form line ray AB), carried first along [10] the ray AB, then along the ray BA, always forward to infinity: the path CD of C will be the line L of CM.

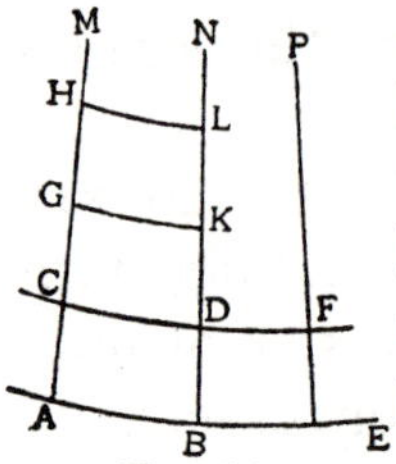

FIG. 16.

For let D be any point in line CD (called later L'), let DN be ‖ CM, and B the point of L falling on the straight DN. We shall have BN $\doteq$ AM, and sect AC=sect BD, and so DN $\doteq$ CM, consequently D in L'. But if D in L' and DN ‖ CM, and B the point of L on the straight DN; we shall have AM $\doteq$ BN and CM $\doteq$ DN, whence manifestly sect BD=sect AC, and D will fall on the path of the point C, and L′ and the line CD are the same. Such an L′ is designated by L′∥L.

§ 23. If the L form line CDF ∥ ABE (§ 22), and AB=BE, and the rays AM, BN, EP are axes; manifestly CD=DF; and if any three points A, B, E are of line AB, and AB=n.CD, we shall also have AE=n.CF; and so (manifestly even for AB, AE, DC incommensurable), AB:CD=AE:CF, and AB:CD is *independent of AB, and completely determined by AC.*

This ratio AB:CD is designated by the capital letter (as X) corresponding to the small letter (as x) by which we represent the sect AC.

§ 24. Whatever be x and y; (§ 23), $Y=X^{\frac{y}{x}}$.

For, one of the quantities x, y is a multiple of the the other (e. g. y of x), or it is not.

If $y=$n.x, take $x=$AC=CG=GH=&c., until we get AH=y.

Moreover, take CD ∥ GK ∥ HL.

We have ((§ 23) X=AB:CD=CD:GK=GK: HL; and so

$$\frac{AB}{HL}=\left[\frac{AB}{CD}\right]^{n}$$

or $Y=X^{n}=X^{\frac{y}{x}}$.

If x, y are multiples of i, suppose $x=mi$, and $y=ni$; (by the preceding) $X=I^{m}$, $Y=I^{n}$, consequently

$$Y=X^{\frac{n}{m}}=X^{\frac{y}{x}}.$$

The same is easily extended to the case of the incommensurability of x and y.

But if q=y−x, manifestly Q=Y:X.

It is also manifest that in Σ, for any x, we have X=1, but in S is X>1, and for any AB and ABE there is such a CDF ||| AB, that CDF =AB, whence AMBN≅AMEP, though the first be any multiple of the second; which indeed is singular, but evidently does not prove the absurdity of S. [II]

§ 25. *In any rectilineal triangle, the circles with radii equal to its sides are as the sines of the opposite angles.*

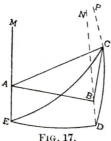

FIG. 17.

For take ∠ABC=rt.∠, and AM ⊥ BAC, and BN and CP || AM; we shall have CAB ⊥ AMBN, and so (since CB ⊥ BA), CB⊥ AMBN, consequently CPBN ⊥ AMBN.

Suppose the F of ray CP cuts the straights BN, AM respectively in D and E, and the bands CPBN, CPAM, BNAM along the L form lines CD, CE, DE. Then (§ 20) ∠CDE= the angle of NDC, NDE, and so =rt.∠; and by like reasoning ∠CED=∠CAB. But (by § 21) in the L line △ CDE (supposing always here the radius =1),

EC:DC=1:sin DEC=1:sin CAB.

Also (by § 21)

EC:DC=⊙EC:⊙DC (in F)=⊙AC:⊙BC (§ 18);

and so is also

⊙AC:⊙BC=1:sin CAB;

whence the theorem is evident for any triangle.

§ **26.** *In any spherical triangle, the sines of the sides are as the sines of the angles opposite.*

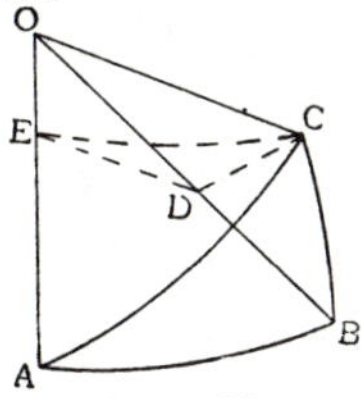

FIG. 18.

For take ∠ABC=rt.∠, and CED ⊥ to the radius OA of the sphere. We shall have CED ⊥ AOB, and (since also BOC ⊥ BOA), CD ⊥ OB. But in the triangles CEO, CDO (by § 25) ⊙EC:⊙OC:⊙DC=sin COE : 1 : sin COD=sin AC : 1 : sin BC; meanwhile also (§ 25) ⊙EC : ⊙DC=sin CDE : sin CED. Therefore, sin AC : sin BC=sin CDE : sin CED; but CDE= rt.∠=CBA, and CED=CAB. Consequently

sin AC : sin BC=1 : sin A.

Spherical trigonometry, flowing from this, is thus established independently of Axiom XI.

§ **27.** If AC and BD are ⊥ AB, and CAB is carried along the straight AB; we shall have, designating by CD the path of the point C,

CD : AB=sin *u* : sin *v*.

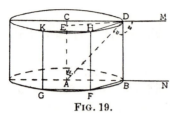

FIG. 19.

For take DE ⊥ CA; in the triangles ADE, ADB (by § 25)

⊙ED : ⊙AD : ⊙AB= sin u : 1 : sin v.

In revolving BACD about AC, B describes ⊙AB, and D describes ⊙ED; and designate here by s⊙CD the path of the said CD. Moreover, let there be any [12] polygon BFG... inscribed in ⊙AB.

Passing through all the sides BF, FG, &c., planes ⊥ to ⊙AB we form also a polygonal figure of the same number of sides in s⊙CD, and we may demonstrate, as in § 23, that CD : AB =DH : BF=HK : FG, &c., and so

DH+HK &c. : BF+FG &c. : =CD : AB.

If each of the sides BF, FG... approaches the limit zero, manifestly

$$BF+FG+\ldots \doteq ⊙AB \quad \text{and}$$
$$DH+HK+\ldots \doteq ⊙ED.$$

Therefore also ⊙ED : ⊙AB=CD : AB. But we had ⊙ED : ⊙AB=sin u : sin v. Consequently

CD : AB=sin u : sin v.

If AC goes away from BD to infinity, CD : AB, and so also sin u : sin v remains *constant;* but $u \doteq$ rt. ∠ (§ 1), and if DM ∥ BN, $v \doteq z$; whence CD : AB=1 : sin z.

The path called CD will be denoted by CD ||| AB.

§ **28.** If BN || ≏ AM, and C in ray AM, and AC=x: we shall have (§ 23)

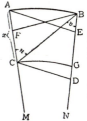

X=sin u : sin v.

For if CD and AE are ⊥ BN, and BF ⊥ AM; we shall have (as in § 27)

⊙BF : ⊙DC=sin u : sin v.

But evidently BF=AE: therefore

⊙EA : ⊙CD=sin u : sin v.

Fig. 20.

But in the F form surfaces of AM and CM (cutting AMBN in AB and CG) (by § 21)

⊙EA : ⊙DC=AB : CG=X.

Therefore also

X=sin u : sin v.

§ **29.** If ∠BAM=rt.∠, and sect AB=y, and BN || AM, we shall have in S

Y=cotan ½ u.

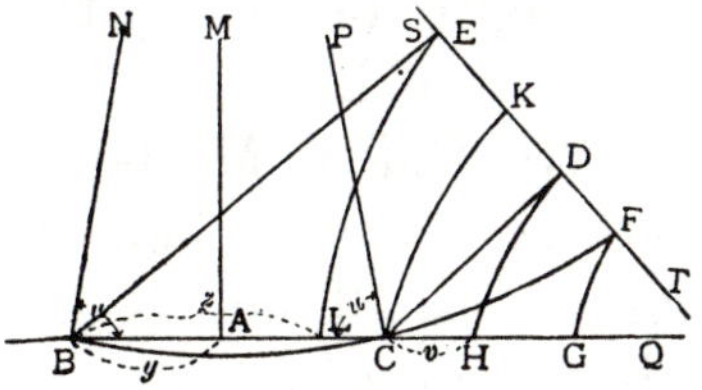

For, if sect AB= sect AC, and CP || AM (and so BN || ≏ CP), and ∠PCD= ∠QCD; there is given (§ 19) DS ⊥ ray CD, so that DS || CP, and so (§ 1) DT || CQ. Moreover, if BE ⊥ ray DS, then (§ 7) DS || BN, and so (§ 6)

Fig. 21.

BN $\|$ ES, and (since DT $\|$ CG) BQ $\|$ ET; consequently (§1) ∠EBN=∠EBQ. Let BCF be an L-line of BN, and FG, DH, CK, EL, L form lines of FT, DT, CQ and ET; evidently (§22) HG=DF=DK=HC; therefore,

$$CG=2CH=2v.$$

Likewise it is evident BG=2BL=2z.

But BC=BG−CG; wherefore $y=z-v$, and so (§24) Y=Z : V.

Finally (§28)

$$Z=1 : \sin \tfrac{1}{2} u,$$
$$\text{and } V=1 : \sin (\text{rt.}\angle - \tfrac{1}{2} u),$$

consequently Y=cotan $\tfrac{1}{2} u$.

§ 30. However, it is easy to see (by § 25) [13] that the solution of the problem of Plane

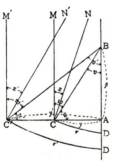

Fig. 22.

Trigonometry, in S, requires the expression of the circle in terms of the radius; but this can by obtained by the rectification of L.

Let AB, CM, C′M′ be ⊥ ray AC, and B anywhere in ray AB; we shall have (§ 25)

$$\sin u : \sin v=\odot p : \odot y,$$
$$\text{and } \sin u' : \sin v' =\odot p' : \odot y';$$

and so $\dfrac{\sin u}{\sin v}.\odot y=\dfrac{\sin u'}{\sin v'}.\odot y'.$

But (by § 27) $\sin v : \sin v' = \cos u : \cos u'$;

consequently $\dfrac{\sin u}{\cos u} . \odot y = \dfrac{\sin u'}{\cos u'} . \odot y'$;

or $\odot y : \odot y' = \tan u' : \tan u = \tan w : \tan w'$.

Moreover, take CN and C'N' ∥ AB, and CD, C'D' L-form lines ⊥ straight AB; we shall have also (§21)

$$\odot y : \odot y' = r : r', \text{ and so}$$
$$r : r' = \tan w : \tan w'.$$

Now let p beginning from A increase to infinity; then $w \overset{\cdot}{=} z$, and $w' \overset{\cdot}{=} z'$, whence also $r : r' = \tan z : \tan z'$.

Designate by i the *constant*

$$r : \tan z \;(independent \text{ of } r);$$

whilst $y \overset{\cdot}{=} 0$,

$$\frac{r}{y} = \frac{i \tan z}{y} \overset{\cdot}{=} 1, \text{ and so}$$

$\dfrac{y}{\tan z} \overset{\cdot}{=} i$.   From § 29, $\tan z = \frac{1}{2}\,(Y - Y^{-1})$;

therefore $\dfrac{2y}{Y - Y^{-1}} \overset{\cdot}{=} i$,

or (§ 24) $\dfrac{2y . I^{\frac{y}{i}}}{I^{\frac{2y}{i}} - 1} \overset{\cdot}{=} i$.

But we know the limit of this expression (where $y \overset{\cdot}{=} 0$) is

$$\frac{i}{\text{nat. log I}}.\quad \text{Therefore}$$

$$\frac{i}{\text{nat. log I}}=i,\ \text{and}$$
$$I=e=2.7182818\ldots,$$

which noted quantity shines forth here also.

If obviously henceforth i denote that sect of which the $I=e$, we shall have

$$r=i \tan z.$$

But (§ 21) $\odot y=2\pi r;$ therefore

$$\odot y=2\pi i \tan z=\pi i\ (Y-Y^{-1})=\pi i \left(e^{\frac{y}{i}}-e^{\frac{-y}{i}} \right)$$
$$=\frac{\pi y}{\text{nat. log Y}}(Y-Y^{-1})\ \text{(by § 24).}$$

§ **31.** For the trigonometric solution of all right-angled rectilineal *triangles* (whence the resolution of all *triangles* is easy ., in S, three [14] equations suffice : indeed (*a*, *b* denoting the sides, *c* the hypothenuse, and *α*, *β* the angles opposite the sides) an equation expressing the relation

1st, between *a*, *c*, *α*;

2d, between *a*, *α*, *β*;

3d, between *a*, *b*, *c*;

of course from these equations emerge three others by elimination.

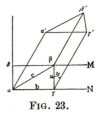

Fig. 23.

From §§ 25 and 30

$$1 : \sin \alpha=(C-C^{-1}) : (A-A^{-1})=$$
$$=\left(e^{\frac{c}{i}}-e^{\frac{-c}{i}} \right) : \left(e^{\frac{a}{i}}-e^{\frac{-a}{i}} \right) \text{ (equation for } c, a \text{ and } \alpha\text{).}$$

II.   From § 27 follows (if βM ‖ γN)
$$\cos a : \sin \beta = 1 : \sin u,\; \text{but from § 29}$$
$$1 : \sin u = \tfrac{1}{2}(A + A^{-1});$$
therefore $\cos a \sin \beta = \tfrac{1}{2}(A+A^{-1}) = \tfrac{1}{2}\left(e^{\frac{a}{i}} + e^{\frac{-a}{i}}\right)$
(equation for a, β and a).

III.   If $aa' \perp \beta a\gamma$, and $\beta\beta'$ and $\gamma\gamma' \| aa'$ (§ 27),
and $\beta'a'\gamma' \perp aa'$; manifestly (as in § 27)
$$\frac{\beta\beta'}{\gamma\gamma'} = \frac{1}{\sin u} = \tfrac{1}{2}(A+A^{-1});$$

$$\frac{\gamma\gamma'}{aa'} = \tfrac{1}{2}(B+B^{-1}\;;$$

and $\dfrac{\beta\beta'}{aa'} = \tfrac{1}{2}(C+C^{-1})$; consequently

$$\tfrac{1}{2}(C+C^{-1}) = \tfrac{1}{2}(A+A^{-1}) . \tfrac{1}{2}(B+B^{-1}), \text{ or}$$

$$\left(e^{\frac{c}{i}} + e^{\frac{-c}{i}}\right) = \tfrac{1}{2}\left(e^{\frac{a}{i}}+e^{\frac{-a}{i}}\right)\left(e^{\frac{b}{i}}+e^{\frac{-b}{i}}\right)$$

(equation for a, b and c).

If $\gamma a\delta = $ rt. $\angle$, and $\beta\,\delta \perp a\delta$;
$$\odot c : \odot a = 1 : \sin a, \text{ and}$$
$$\odot c : \odot (d = \beta\delta) = 1 : \cos a,$$
and so (denoting by $\odot x^2$, for any x, the product $\odot x . \odot x$) manifestly
$$\odot a^2 + \odot d^2 = \odot c^2.$$

But (by § 27 and II)
$$\odot d = \odot b . \tfrac{1}{2}(A+A^{-1}), \text{ consequently}$$

$$\left(e^{\frac{c}{i}} - e^{\frac{-c}{i}}\right)^2 = \tfrac{1}{4}\left(e^{\frac{a}{i}}+e^{\frac{-a}{i}}\right)^2 . \left(e^{\frac{b}{i}}-e^{\frac{-b}{i}}\right)^2 + \left(e^{\frac{a}{i}}-e^{\frac{-a}{i}}\right)^2$$

another equation for a, b and c (the second

member of which may be easily reduced to a form *symmetric* or *invariable*).   [15]

Finally, from

$\dfrac{\cos a}{\sin \beta} = \tfrac{1}{2}(A + A^{-1})$, and $\dfrac{\cos \beta}{\sin a} = \tfrac{1}{2}(B + B^{-1})$, we get (by III)

$$\cot a \cot \beta = \tfrac{1}{2}\left[e^{\frac{c}{i}} + e^{\frac{-c}{i}} \right]$$

(equation for a, β, and c.

§ **32.** It still remains to show briefly the mode of resolving *problems* in S, which being accomplished (through the more obvious examples), finally will be candidly said what this theory shows.

I.   Take AB a line in a plane, and $y = f(x)$

FIG. 24.

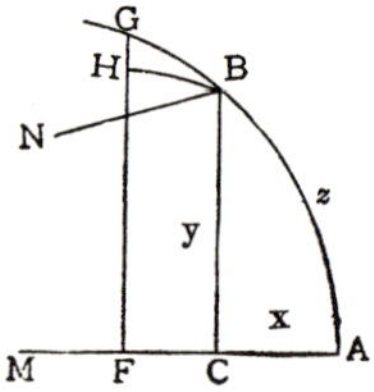

its equation in rectangular coordinates, call dz any increment of z, and respectively dx, dy, du the increments of x, of y, and of the area u, corresponding to this dz; take BH $|||$ CF, and express (from § 31) $\dfrac{BH}{dx}$ by means of y, and seek the *limit* of $\dfrac{dy}{dx}$ when dx tends towards the limit zero (which is understood where a limit of this sort is sought): then will become known also the limit of $\dfrac{dy}{BH}$, and so tan HBG; and (since HBC manifestly is neither $>$ nor $<$, and so $=$rt.$\angle$), the *tangent* at B of BG will be determined by y.

II. It can be demonstrated

$$\frac{dz^2}{dy^2+\overline{\overline{\text{BH}}}^2}{=}1.$$

Hence is found the *limit* of $\frac{dz}{dx}$, and thence, by integration, z (expressed in terms of x.

And of any line *given in the concrete,* the equation in S can be found; e. g., of L. For if ray AM be the axis of L; then any ray CB from ray AM cuts L [since (by § 19) any straight from A except the straight AM will cut L]; but (if BN is axis)

X$=1$:sin CBN (§ 28),
and Y$=$cotan ½ CBN (§ 29), whence
Y$=$X$+\sqrt{\text{X}^2-1.}$

or $\qquad e^{\frac{y}{i}}=e^{\frac{x}{i}}+\sqrt{e^{\frac{2x}{i}}-1,}$ [16]

the equation sought.

Hence we get

$$\frac{dy}{dx}{=}\text{X}(\text{X}^2-1)^{-\frac{1}{2}};$$

and $\dfrac{\text{BH}}{dx}{=}1:\sin$ CBN$=$X; and so

$$\frac{dy}{\text{BH}}{=}(\text{X}^2-1)^{-\frac{1}{2}};$$

$$1+\frac{dy^2}{\mathrm{BH}^2}\dot{=}\mathrm{X}^2(\mathrm{X}^2-1)^{-1},$$

$$\frac{dz^2}{\mathrm{BH}^2}\dot{=}\mathrm{X}^2(\mathrm{X}^2-1)^{-1},$$

and $\dfrac{dz}{\mathrm{BH}}\dot{=}\mathrm{X}(\mathrm{X}^2-1)^{-\frac{1}{2}}$, and

$\dfrac{dz}{dx}\dot{=}\mathrm{X}^2(\mathrm{X}^2-1)^{-\frac{1}{2}}$, whence, by integration, we get (as in § 30)

$$z=i(\mathrm{X}^2-1)^{\frac{1}{2}}=i\cot\mathrm{CBN}.$$

III.   Manfestly

$$\frac{du}{dx}\dot{=}\frac{\mathrm{HFCBH}}{dx},$$

which (unless given in y) now first is to be expressed in terms of $y;$ whence we get u by integrating.

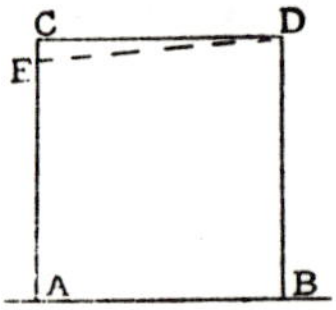

FIG. 25.

If AB$=p$, AC$=q$, CD$=r$, and CABDC$=s;$ we might show (as in II) that

$$\frac{ds}{dq}\dot{=}r,\ \text{which}\ =\tfrac{1}{2}p\left(e^{\frac{q}{i}}-e^{\frac{-q}{i}}\right)$$

and, integrating, $s=\tfrac{1}{2}pi\left[e^{\frac{q}{i}}-e^{\frac{-q}{i}}\right]$

This can also be deduced apart from integration.

For example, the equation of the circle (from § 31, III), of the straight (from § 31, II), of a conic (by what precedes), being expressed, the areas bounded by these lines could also be expressed.

We know, that a surface t, ‖ to a plane figure p (at the distance q), is to p in the ratio of the second powers of homologous lines, or as

$$\tfrac{1}{4}\left(e^{\frac{q}{i}}-e^{\frac{-q}{i}}\right)^{2}:1.$$

It is easy to see, moreover, that the calculation of volume, treated in the same manner, requires two integrations (since the differential itself here is determined only by integration); and before all must be investigated the [17] volume contained between p and t, and the aggregate of all the straights $\perp p$ and joining the boundaries of p and t.

We find for the volume of this solid (whether by integration or without it)

$$\tfrac{1}{8}pi\left(e^{\frac{2q}{i}}-e^{\frac{-2q}{i}}\right)+\tfrac{1}{2}pq.$$

The surfaces of bodies may also be determined in S, as well as the *curvatures,* the *involutes*, and *evolutes* of any lines, etc.

As to curvature; this in S either is the curvature of L, or is determined either by the radius of a circle, or by the *distance* to a straight from the curve ‖ to this straight; since from what precedes, it may easily be shown, that in a plane there are no uniform lines other than L-lines, circles and curves ‖ to a straight.

IV. For the circle (as in III) $\dfrac{d \text{ area} \odot x}{dx} =$ $\odot x$, whence (by § 29), integrating,

$$\text{area} \odot x = \pi i^2 \left[e^{\frac{x}{i}} - 2 + e^{\frac{-x}{i}} \right].$$

V. For the area CABDC$=u$ (inclosed by an L form line AB$=r$, the ‖ to this, CD$=y$, and the sects AC$=$BD$=x$) $\dfrac{du}{dx} = y$; and (§ 24) $y = re^{\frac{-x}{i}}$, and so (integrating) $u = ri \left[1 - e^{\frac{-x}{i}} \right]$.

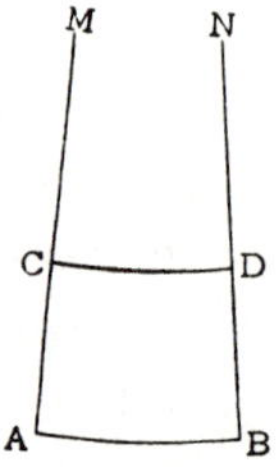

If x increases to infinity, then, in S, $e^{\frac{-x}{i}} = 0$, and so $u = ri$. By the *size* of MABN, in future this limit is understood.

FIG. 26.

In like manner is found, if p is a figure on F, the space included by p and the aggregate of axes drawn from the boundaries of p is equal to ½pi.

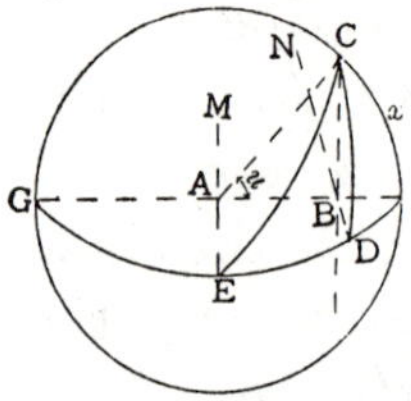

FIG. 27.

VI. If the angle at the center of a segment z of a sphere is $2u$, and a great circle is p, and x the arc FC (of the angle u); (§ 25)

$$1 : \sin u = p : \odot BC,$$

and hence $\odot BC = p \sin u.$ [18]

Meanwhile $x = \dfrac{pu}{2\pi}$, and $dx = \dfrac{pdu}{2\pi}.$

Moreover, $\dfrac{dz}{dx} \doteq \odot BC$, and hence

$$\dfrac{dz}{du} \doteq \dfrac{p^2}{2\pi} \sin u, \text{ whence (integrating)}$$

$$z = \dfrac{\text{ver sin } u}{2\pi} p^2.$$

The F may be conceived on which P falls (passing through the middle F of the segment); through AF and AC the planes FEM, CEM are placed, perpendicular to F and cutting F along FEG and CE; and consider the L form CD (from C ⊥ to FEG), and the L form CF; (§ 20) CEF$=u$, and (§ 21)

$$\dfrac{FD}{p} = \dfrac{\text{ver sin } u}{2\pi}, \text{ and so } z = FD.p.$$

But (§ 21) $p = \pi.FGD$; therefore

$z = \pi.FD.FDG.$ But (§ 21)

FD.FDG=FC.FC; consequently

$z = \pi.FC.FC = $ area $\odot FC$, in F.

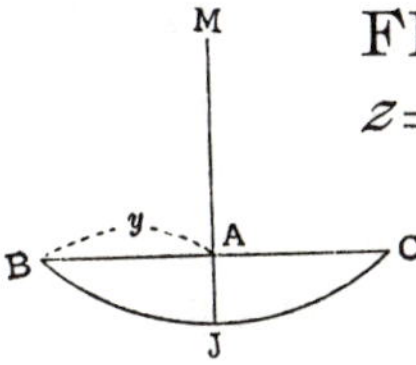

Now let BJ=CJ=r; (§ 30) $2r = i(Y - Y^{-1})$, and so (§ 21) area $\odot 2r$ (in F) $= \pi i^2(Y - Y^{-1})^2$.

FIG. 28.   Also (*IV*)

area $\odot 2y = \pi i^2(Y^2 - 2 + Y^{-2})$;

therefore, area $\odot 2r$ (in F) =area $\odot 2y$, and so *the surface z of a segment of a sphere is equal to the surface of the circle described with the chord* FC *as a radius.*

Hence the whole surface of the sphere

$$= \text{area} \odot FG = FDG.p = \frac{p^2}{\pi},$$

and the surfaces of spheres are to each other as the second powers of their great circles.

VII. In like manner, in S, the volume of the sphere of radius x is found

$$= \tfrac{1}{2}\pi i^3 (X^2 - X^{-2}) - 2\pi i^2 x;$$

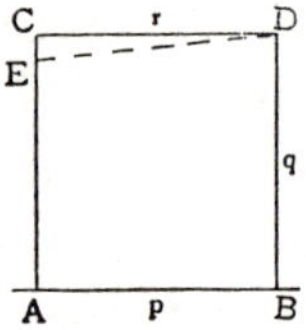

the surface generated by the revolution of the line CD about AB

$$= \tfrac{1}{2}\pi ip(Q^2 - Q^{-2}),$$

and the body described by CABDC

$$= \tfrac{1}{4}\pi i^2 p(Q - Q^{-1})^2.$$

FIG. 29.

But in what manner all things treated from (IV) even to here, also may be reached apart from integration, for the sake of brevity is suppressed.

It can be demonstrated that *the limit of every expression containing the letter i* (and so resting upon the hypothesis that i is given), [19] *when i increases to infinity, expresses the quantity simply for Σ* (and so for the hypothesis of no i), if indeed the equations do not become identical.

But beware lest you understand to be supposed, that the system itself may be varied (for it is entirely determined in itself and by itself); but only *the hypothesis,* which may be done successively, as long as we are not conducted to an absurdity. *Supposing* therefore that, in *such* an expression, the letter *i*, in case S is reality, designates that unique quantity whose I=*e;* but if Σ is actual, the said limit is supposed to be taken in place of the expression: manifestly *all the expressions originating from the hypothesis of the reality of S (in this sense) will be true absolutely, although it be completely unknown whether or not Σ is reality*

So e. g. from the expression obtained in § 30 easily (and as well by aid of differentiation as apart from it) emerges the known value in Σ,

$$\odot x = 2\pi x;$$

from I (§ 31) suitably treated, follows

$$1 : \sin a = c : a;$$

but from II

$$\frac{\cos a}{\sin \beta} = 1, \text{ and so}$$

$$a + \beta = \text{rt.} \angle;$$

the first equation in III becomes identical, and so is true in Σ, although it there determines nothing; but from the second follows

$$c^2 = a^2 + b^2.$$

These are the known fundamental equations of plane trigonometry in Σ.

Moreover, we find (from § 32) in Σ, the area and the volume in III each $=pq;$ from IV

$$\text{area} \odot x = \pi x^2;$$

(from VII) the globe of radius x

$$= \tfrac{4}{3}\pi x^3, \text{ etc.}$$

The theorems enunciated at the end of VI are manifestly *true unconditionally*.

§ **33.** It still remains to set forth (as promised in § 32) what this theory means.

I.   Whether Σ or some one S is reality, remains undecided.

II.   All things deduced from the hypothesis of the falsity of Axiom *XI* (always to be understood in the sense of § 32) are *absolutely true,* and so in this sense, *depend upon no hypothesis.*

There is therefore *a plane trigonometry a priori, in which the system alone really re-* [20] *mains unknown;* and so where remain unknown solely the *absolute* magnitudes in the expressions, but where a *single* known case would manifestly fix the whole system. But spherical trigonometry is established absolutely in § 26.

(And we have, on F, a geometry wholly analogous to the plane geometry of Σ.)

III.   If it were agreed that Σ exists, nothing more would be unknown in this respect; but

if it were *established* that Σ *does not exist*,
then (§ 31), (e. g.) from the sides *x, y,* and the
rectilineal angle they include being given in a
special case, manifestly it would be impossible
in itself and by itself to solve absolutely the
triangle, that is, to determine *a priori* the
other angles and *the ratio of the third side* to
the two given; unless X, Y were determined,
for which it would be necessary to have in
concrete form a certain sect *a* whose A was
known; and then *i* would be *the natural unit
for length* (just as *e* is the base of *natural*
logarithms).

   If the existence of this *i* is determined, it
will be evident how it could be constructed,
at least very exactly, for practical use.

   IV.   In the sense explained (I and II), it is
evident that all things in space can be solved
by the modern analytic method (within just
limits strongly to be praised).

   V.   Finally, to friendly readers will not be
unacceptable; that for that case wherein not Σ
but S is reality, a rectilineal figure is constructed equivalent to a circle.

   § **34.**   Through D we may draw DM ‖ AN in
the following manner.   From D drop DB⊥AN;
from any point A of the straight AB erect AC
⊥ AN (in DBA), and let fall DC⊥AC.   We

will have (§ 27) ⊙CD : ⊙AB=1 : sin *z*, pro-

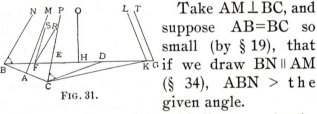

vided that DM ‖ BN. But sin *z* is not >1; and so AB is not >DC. Therefore a quadrant described from the center A in BAC, with a radius =DC, will have a point B or O in common with ray BD. In the first case, manifestly *z*=rt.∠; but in the second case (§ 25)

(⊙AO=⊙CD) : ⊙AB=1 : sin AOB,

and so *z*=AOB.

If therefore we take *z*=AOB, then DM will be ‖ BN.·

§ **35.** If S were reality; we may, as follows, draw a straight ⊥ to one arm of an acute angle, [21] which is ‖ to the other.

Take AM ⊥ BC, and suppose AB=BC so small (by § 19), that if we draw BN ‖ AM (§ 34), ABN > the given angle.

FIG. 31.

Moreover draw CP ‖ AM (§ 34); and take NBG and PCD each equal to the given angle; rays BG and CD will cut; for if ray BG (falling *by construction* within NBC) cuts ray CP in E; we shall have (since BN ≏ CP), ∠EBC< ∠ECB, and so EC<EB. Take EF=EC, EFR

=ECD, and FS ‖ EP; then FS will fall within
BFR. For since BN ‖ CP, and so BN ‖ EP,
and BN ‖ FS; we shall have (§ 14)

$$\angle FBN + \angle BFS < (st.\angle = FBN + BFR);$$

therefore, BFS < BFR. Consequently, ray FR
cuts ray EP, and so ray CD also cuts ray EG
in some point D. Take now DG=DC and
DGT=DCP=GBN; we shall have (since CD ≏
GD) BN ≏ GT ≏ CP. Let K (§ 19) be the point
of the L-form line of BN falling in the ray BG,
and KL the axis; we shall have BN ≏ KL,
and so BKL=BGT=DCP; but also KL ≏ CP:
therefore manifestly K fall on G, and GT ‖ BN.
But if HO bisects ⊥ BG, we shall have constructed HO ‖ BN.

§ 36. Having given the ray CP and the
plane MAB, take CB ⊥ the
plane MAB, BN (in plane
BCP) ⊥ BC, and CQ ‖ BN
(§ 34); the intersection of ray
CP (if this ray falls within
BCQ) with ray BN (in the
plane CBN), and so with the
plane MAB is found. And if we are given
the two planes PCQ, MAB, and we have CB
⊥ to plane MAB, CR ⊥ plane PCQ; and (in
plane BCR) BN⊥BC, CS⊥CR, BN will fall
in plane MAB, and CS in plane PCQ; and the

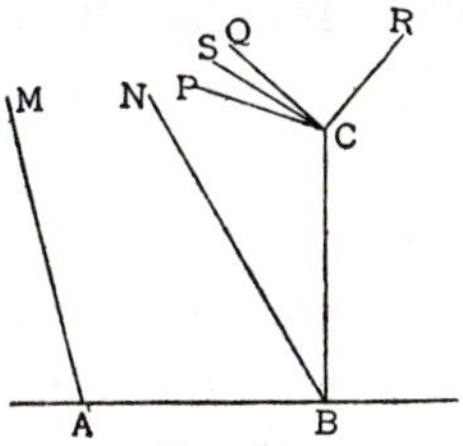

FIG. 32.

intersection of the straight BN with the
straight CS (if there is one) having been found,
the perpendicular drawn through this intersection, in PCQ, to the straight CS will manifestly be the intersection of plane MAB and
plane PCQ.

§ 37. On the straight AM ∥ BN, is found such

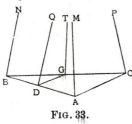

an A, that AM ⇌ BN. If (by [22]
§ 34) we construct outside
of the plane NBM, GT ∥
BN, and make BG⊥GT,
GC=GB, and CP∥GT;
and so place the hemi-
plane TGD that it makes

FIG. 33.

with hemi-plane TGB an angle equal to that
which hemi-plane PCA makes with hemi-plane
PCB; and is sought (by § 36) the intersection
straight DQ of hemi-plane TGD with hemi-
plane NBD; and BA is made ⊥ DQ.

We shall have indeed, on account of the sim-
ilitude of the triangles of L lines produced on
the F of BN (§ 21), manifestly DB=DA, and
AM ⇌ BN.

Hence easily appears (L-lines being given by
their extremities alone) we may also find a
fourth proportional, or a mean proportional,
and execute in this way in F, apart from Axiom XI, all the geometric constructions made

on the plane in *Σ*. Thus e. g. a perigon can be geometrically divided into any special number of equal parts, if it is permitted to make this special partition in *Σ*.

§ **38.** If we construct (by § 37) for example,

NBQ=⅓ rt.∠, and make (by § 35), in S, AM⊥ray BQ and ∥ BN, and determine (by § 37) IM≙BN; we shall have, if IA

Fig. 34.   =*x*, (§ 28), X=1 : sin ⅓ rt. ∠=2, and *x* will be constructed *geometrically*.

And NBQ may be so computed, that IA differs from *i* less than by anything given, which happens for sin NBQ=¹/*e*.

§ **39.** If (in a plane) PQ and ST are ∥ to the straight MN (§ 27), and AB, CD are equal perpendiculars to MN; manifestly △DEC≅

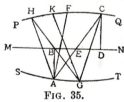

△BEA; and so the angles (perhaps mixtilinear) ECP, EAT will fit, and EC=EA. If, moreover, CF=AG, then △ACF≅△CAG, and each is half of the *quadrilateral*

Fig. 35.

FAGC.

If FAGC, HAGK are two quadrilaterals of this sort on AG, between PQ and ST; their equivalence (as in Euclid) is evident, as also the equivalence of the triangles AGC, AGH, standing on the same AG, and having their vertices on the line PQ. Moreover, ACF= CAG, GCQ=CGA, and ACF+ACG+GCQ= st. ∠ (§ 32); and so also CAG+ACG+CGA= [23] st. ∠; therefore, in any triangle ACG of this sort, the sum of the three angles =st. ∠. But whether the straight AG may have fallen upon AG (which ⫴ MN), or not; *the equivalence* of the rectilineal triangles AGC, AGH, as well of themselves, *as of the sums of their angles,* is evident.

§ **40**. *Equivalent triangles* ABC, ABD, (henceforth rectilineal), *having one side equal, have the sums of their angles equal.*

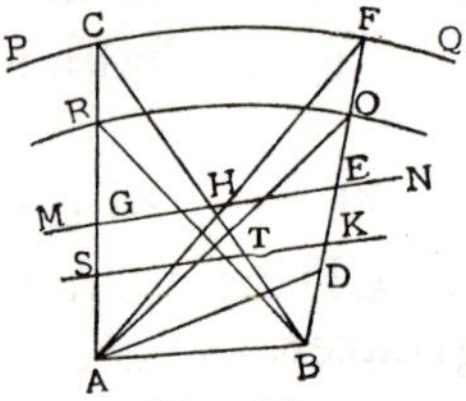

FIG. 36.

For let MN bisect AC and BC, and take (through C) PQ ⫴ MN; the point D will fall on line PQ.

For, if ray BD cuts the straight MN in the point E, and so (§ 39) the line PQ at the distance EF=EB; we shall have △ABC=△ABF, and so also △ABD=△ABF, whence D falls at F.

But if ray BD has not cut the straight MN, let C be the point, where the perpendicular bisecting the straight AB cuts the line PQ, and let GS=HT, so, that the line ST meets the ray BD prolonged in a certain K (which it is evident can be made in a way like as in § 4); moreover take SR=SA, RO ⫴ ST, and O the intersection of ray BK with RO; then △ABR =△ABO (§ 39), and so △ABC>△ABD (contra hyp.).

§ **41.** *Equivalent   triangles   ABC, DEF have the sums of their triangles equal.*

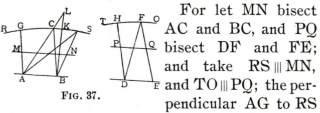

FIG. 37.

For let MN bisect AC and BC, and PQ bisect DF and FE; and take RS ⫴ MN, and TO ⫴ PQ; the perpendicular AG to RS will equal the perpendicular DH to TO, or one for example DH will be the greater.

In each case, the ⊙DF, from center A, has with line-ray GS some point K in common, and (§ 39) △ABK=△ABC=△DEF. But the △AKB (by § 40) has the same angle-sum as △DFE, and (by § 39) as △ABC. Therefore also the triangles ABC, DEF have each the same angle-sum.

In S the inverse of this theorem is true.

For take ABC, DEF two triangles having equal angle-sums, and △BAL=△DEF; these will have (by what precedes) equal angle-sums, and so also will △ABC and △ABL, and hence manifestly

$$BCL+BLC+CBL=\text{st.}\angle.$$

However (by § 31), the angle-sum of any tri- [24] angle, in S, is <st.∠.

Therefore L falls on C.

§ 42. Let u be the supplement of the angle-sum of the △ABC, but v of △DEF; then is △ABC:△DEF=u : v.

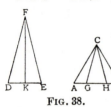

Fig. 38.

For if p be the area of each of the triangles ACG, GCH, HCB, DFK, KFE; and △ABC=$m.p$, and △DEF= $n.p$; and s the angle-sum of any triangle equivalent to p; manifestly

$$\text{st.}\angle-u=m.s-(m-1)\text{st.}\angle=\text{st.}\angle-m(\text{st.}\angle-s);$$

and $u=m(\text{st.}\angle-s)$; and in like manner $v= n(\text{st.}\angle-s)$.

Therefore △ABC : △DEF=m : n=u:v.

It is evidently also easily extended to the case of the incommensurability of the triangles ABC, DEF.

In the same way is demonstrated that triangles on a sphere are as the *excesses* of the sums of their angles above a st.<.

If two angles of the spherical △ are right, the third z will be the said *excess*. But (a great circle being called p) this △ is manifestly

$$=\frac{z}{2\pi}\frac{p^2}{2\pi} \text{ (§ 32, VI)};$$

consequently, any triangle of whose angles the excess is z, is

$$=\frac{zp^2}{4\pi^2}.$$

§ **43.** Now, in S, the area of a rectilineal △ is expressed by means of the sum of its angles.

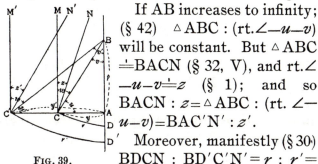

FIG. 39.

If AB increases to infinity; (§ 42) △ABC : (rt.∠—u—v) will be constant. But △ABC $=$BACN (§ 32, V), and rt.∠ —u—$v=z$ (§ 1);   and so BACN : $z=$△ABC : (rt. ∠— u—v)=BAC′N′ : z'. Moreover, manifestly (§ 30) BDCN : BD′C′N′$=r$: $r'=$ tan z : tan z'.

But for $y'=$0, we have

$$\frac{BD'C'N'}{BAC'N'}=1, \text{ and also } \frac{\tan z'}{z'}=1;$$

consequently,

$$BDCN : BACN=\tan z : z.$$

But (§ 32)

$$BDCN=r.i=i^2\tan z;$$

therefore, $BACN=z.i^2$.

Designating henceforth, for brevity, any triangle the supplement of whose angle-sum is z by $\triangle$, we will therefore have $\triangle = z.i^2$.

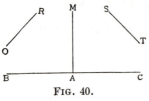

FIG. 40.

Hence it readily flows that, if OR∥AM and RO∥AB, the *area* comprehended between the straights OR, ST, BC [25] (which is manifestly the absolute limit of the area of rectilineal triangles increasing without bound, or of $\triangle$ for $z=$st.$\angle$), is $=\pi i^2=$ area $\odot i$, in F.

This limit being denoted by $\square$, moreover (by § 30) $\pi r^2=\tan^2 z.\square=$ area $\odot r$ in F (§ 21)= area $\odot s$ (by §32, VI) if the chord CD is called s.

If now, bisecting at right angles the given radius s of the circle in a plane (or the L form radius of the circle in F), we construct (by § 34) DB∥⌐CN; by dropping CA ⊥ DB, and erecting CM ⊥ CA, we shall

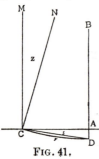

FIG. 41.

get $z;$ whence (by § 37), assuming at pleasure an L form radius for unity, *tan^{2}z can be determined geometrically by means of two uniform lines of the same curvature* (which, their extremities alone being given and their axes constructed, manifestly may be compared like straights, and in this respect considered equivalent to straights).

Moreover, a quadrilateral, ex. gr. regular $= \square$ is constructed as follows:

FIG. 42.

Take ABC=rt.$\angle$, BAC=$\frac{1}{2}$ rt. $\angle$, ACB=$\frac{1}{4}$ rt. $\angle$, and BC=x.

By mere square roots, X (from § 31, II) can be expressed and (by § 37) constructed; and having X (by § 38 or also §§ 29 and 35), x itself can be determined. And octuple $\triangle$ ABC is manifestly $= \square$, and by this *a plane circle of radius s is geometrically squared by means of a rectilinear figure and uniform lines of the same species* (equivalent to straights as to comparison *inter se*); *but an F form circle is planified in the same manner: and we have either the Axiom* XI *of Euclid true or the geometric quadrature of the circle*, although thus far it has remained undecided, which of these two has place in reality.

Whenever $\tan^2 z$ is either a whole number, or a rational fraction, whose denominator (reduced to the simplest form) is either a prime number of the form 2^m+1 (of which is also $2=2^0+1$), or a product of however many prime numbers of this form, of which each (with the exception of 2, which alone may occur any number of times) occurs *only once* as factor, we can, by the theory of polygons of the illustrious Gauss (remarkable invention of our, nay of every age) (and only for such values [26] of z), construct a rectilineal figure $=\tan^2 z \,\square =$ area $\odot s$. For the division of $\square$ (the theorem of § 42 extending easily to any polygons) manifestly requires the partition of a st. $\angle$, which (as can be shown) can be achieved geometrically only under the said condition.

But in all such cases, what precedes conducts easily to the desired end. And any rectilineal figure can be converted geometrically into a regular polygon of n sides, if n falls under the Gaussian form.

It remains, finally (that the thing may be completed in every respect), to demonstrate the impossibility (apart from any supposition), of deciding *a priori*, whether Σ, or some S (and which one) exists. This, however, is reserved for a more suitable occasion.

APPENDIX I.

REMARKS ON THE PRECEDING TREATISE,
BY BOLYAI FARKAS.

[From Vol. II of Tentamen, pp. 380–383.]

Finally it may be permitted to add something appertaining to the author of the *Appendix* in the first volume, who, however, may pardon me if something I have not touched with his acuteness.

The thing consists briefly in this: *the formulas of spherical trigonometry* (demonstrated in the said *Appendix* independently of Euclid's Axiom XI) *coincide with the formulas of plane trigonometry*, *if* (in a way provisionally speaking) *the sides of a spherical triangle are accepted as reals, but of a rectilineal triangle as imaginaries;* so that, as to trigonometric formulas, the plane may be considered as an imaginary sphere, if for real, that is accepted in which sin rt. $\angle = 1$.

Doubtless, of the Euclidean axiom has been said in volume first enough and to spare: for the case if it were not true, is demonstrated (Tom. I. App., p. 13), that there is given a certain i, for which the I there mentioned is $=e$ (the base of natural logarithms), and for this case are established also (*ibidem*, p. 14) the formulas of plane trigonometry, and indeed so, that (by the side of p. 19, ibidem) the formulas are still valid for the case of the verity of the said axiom; indeed if the limits of the values are taken, supposing that $i \overset{\text{\tiny 1}}{=} \infty$; truly the Euclidean system is as if the limit of the anti-Euclidean (for $i \overset{\text{\tiny 1}}{=} \infty$).

Assume for the case of i existing, the unit $=i$, and extend the concepts sine and cosine also to imaginary arcs, so that, p designating an arc whether real or imaginary,

$$\frac{e^{p\sqrt{-1}} + e^{-p\sqrt{-1}}}{2} \text{ is called the}$$

cosine of p, and

$$\frac{e^{p\sqrt{-1}} - e^{-p\sqrt{-1}}}{2\sqrt{-1}} \text{ is called}$$

the *sine* of p (as Tom. I., p. 177).

Hence for q real

$$\frac{e^{q} - e^{-q}}{2\sqrt{-1}} = \frac{e^{-q\sqrt{-1}\cdot\sqrt{-1}} - e^{q\sqrt{-1}\cdot\sqrt{-1}}}{2\sqrt{-1}} = \sin(-q\sqrt{-1})$$

$$= -\sin(q\sqrt{-1}).$$

So $\dfrac{e^{q}+e^{-q}}{2}=\dfrac{e^{-q\sqrt{-1}\cdot\sqrt{-1}}+e^{q\sqrt{-1}\cdot\sqrt{-1}}}{2}=\cos(-q\sqrt{-1})$

$$=\cos(q\sqrt{-1});$$

if of course also in the imaginary circle, the sine of a negative arc is the same as the sine of a positive arc otherwise equal to the first, except that it is negative, and the cosine of a positive arc and of a negative (if otherwise they be equal) the same.

In the said *Appendix*, § 25, is demonstrated absolutely, that is, independently of the said axiom; that, in any rectilineal triangle *the sines of the circles are as the circles of radii equal to the sides opposite.*

Moreover is demonstrated for the case of i existing, that the circle of radius y is

$$=\pi i\left(e^{\frac{y}{i}}-e^{\frac{-y}{i}}\right),\text{ which, for }i=1,\text{ becomes}$$

$$\pi(e^{y}-e^{-y}).$$

Therefore (§ 31 *ibidem*), for a right-angled rectilineal triangle of which the sides are a and b, the hypothenuse c, and the angles opposite to the sides a, b, c are a, β, rt. $\angle$, (for $i=1$), in I,

$$1:\sin a=\pi(e^{c}-e^{-c}):\pi(e^{a}-e^{-a});$$

and so

$$1:\sin a=\frac{e^{c}-e^{-c}}{2\sqrt{-1}}:\frac{e^{a}-e^{-a}}{2\sqrt{-1}}. \qquad \text{Whence } 1:\sin a$$

$$= -\sin\ (c\sqrt{-1}) : -\sin\ (a\sqrt{-1}).$$ And hence

$$1 : \sin\ a = \sin\ (c\sqrt{-1}) : \sin\ (a\sqrt{-1}).$$

In II becomes

$$\cos\ a : \sin\ \beta = \cos\ (a\sqrt{-1}) : 1;$$

in III becomes

$$\cos\ (c\sqrt{-1}) = \cos\ (a\sqrt{-1}).\cos\ (b\sqrt{-1}).$$

These, as all the formulas of plane trigonometry deducible from them, coincide completely with the formulas of spherical trigonometry; except that if, ex. gr., also the sides and the angles opposite them of a right-angled spherical triangle and the hypothenuse bear the same names, the sides of the rectilineal triangle are to be divided by $\sqrt{-1}$ to obtain the formulas for the spherical triangle.

Obviously we get (clearly as Tom., II., p. 252),

from I, $\quad 1 : \sin\ a = \sin\ c : \sin\ a;$

from II, $\quad 1 : \cos\ a = \sin\beta : \cos\ a;$

from III, $\quad \cos\ c = \cos\ a \cos\ b.$

Though it be allowable to pass over other things; yet I have learned that the reader may be offended and impeded by the deduction omitted, (Tom. I., App., p. 19) [in § 32 at end]: it will not be irrelevant to show how, ex. gr., from

$$e^{\frac{c}{i}} + e^{\frac{-c}{i}} = \tfrac{1}{2} \left(e^{\frac{a}{i}} + e^{\frac{-a}{i}} \right) \left(e^{\frac{b}{i}} + e^{\frac{-b}{i}} \right)$$

follows

$$c^2 = a^2 + b^2.$$

(the theorem of Pythagoras for the Euclidean system); probably thus also the author deduced it, and the others also follow in the same manner.

Obviously we have, the powers of e being expressed by series (like Tom. I., p. 168),

$$e^{\frac{k}{i}} = 1 + \frac{k}{i} + \frac{k^2}{2i^2} + \frac{k^3}{2.3.i^3} + \frac{k^4}{2.3.4.i^4} \cdots \cdots ,$$

$$e^{\frac{k}{i}} = 1 - \frac{k}{i} + \frac{k^2}{2i^2} - \frac{k^3}{2.3.i^3} + \frac{k^4}{2.3.4.i^4} \cdots \cdots , \text{ and so}$$

$$e^{\frac{k}{i}} + e^{\frac{-k}{i}} = 2 + \frac{k^2}{i^2} + \frac{k^4}{3.4.i^4} + \frac{k^6}{3.4.5.6.i^6} \cdots \cdots ,$$

$$= 2 + \frac{k^2 + u}{i^2}, \text{ (designating by}$$

$\frac{u}{i^2}$ the sum of all the terms after $\frac{k^2}{i^2}$) ; and we have $u \doteq 0$, while $i \doteq \infty$. For all the terms which follow $\frac{k^2}{i^2}$ are divided by i^2; the first term will be $\frac{k^4}{3.4i^2}$; and any ratio $< \frac{k^2}{i^2}$; and though the ratio everywhere should remain this, the sum would be (Tom. I., p. 131),

$$\frac{k^4}{3.4.i^2} : \left(1 - \frac{k^2}{i^2}\right) = \frac{k^4}{3.4.(i^2 - k^2)},$$

which manifestly $\doteq 0$, while $i \doteq \infty$.

And from

$$e^{\frac{c}{i}}+e^{\frac{-c}{i}}=\tfrac{1}{2}\left(e^{\frac{(a+b)}{i}}+e^{\frac{-(a+b)}{i}}+e^{\frac{a-b}{i}}+e^{\frac{-(a-b)}{i}}\right)$$

follows (for w, v, λ taken like u)

$$2+\frac{c^2+w}{i^2}=\tfrac{1}{2}\left(2+\frac{(a+b)^2+v}{i^2}+2+\frac{(a+b)^2+\lambda}{i^2}\right).$$

And hence

$$c^2=\frac{a^2+2ab+b^2+a^2-2ab+b^2+v+\lambda-w}{2},$$

which $\overset{\cdot}{=}a^2+b^2$.

APPENDIX II.

SOME POINTS IN JOHN BOLYAI'S APPENDIX
COMPARED WITH LOBACHEVSKI,
BY WOLFGANG BOLYAI.

[From *Kurzer Grundriss*, p. 82.]

Lobachevski and the author of the *Appendix* each consider two points A, B, of the sphere-limit, and the corresponding axes ray AM, ray BN (§ 23).

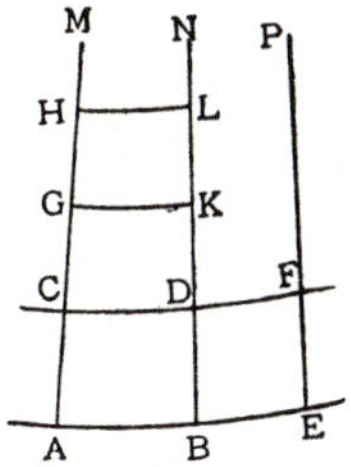

They demonstrate that, if a, β, γ designate the arcs of the circle limit AB, CD, HL, separated by segments of the axis AC=1, AH $=x$, we have

$$\frac{a}{\gamma} = \left(\frac{a}{\beta}\right)^{x}.$$

FIG. 43.

Lobachevski represents the value of $\frac{\gamma}{a}$ by e^{-x}, e having some value >1, dependent on the unit for length that we have chosen, and able to be supposed equal to the Naperian base.

The author of the *Appendix* is led directly to introduce the base of natural logarithms.

If we put $\frac{a}{\beta} = \delta$, and γ, γ' are arcs situated at the distances y, i from a, we shall have

$$\frac{a}{\gamma} = \delta^y = Y, \qquad \frac{a}{\gamma'} = \delta^i = I, \quad \text{whence} \quad Y = I^{\frac{y}{i}}.$$

He demonstrates afterward (§ 29) that, if u is the angle which a straight makes with the perpendicular y to its parallel, we have

$$Y = \cot \tfrac{1}{2}u.$$

Therefore, if we put $z = \frac{\pi}{2} - u$, we have

$$Y = \tan\ (z + \tfrac{1}{2}u) = \frac{\tan\ z + \tan\ \tfrac{1}{2}u}{1 - \tan\ z\ \tan\ \tfrac{1}{2}u},$$

whence we get, having regard to the value of $\tan \tfrac{1}{2}u = Y^{-1}$,

$$\tan\ z = \tfrac{1}{2}\ (Y - Y^{-1}) = \tfrac{1}{2} \left[I^{\frac{y}{i}} - I^{\frac{-y}{i}} \right] \text{ (§ 30)}.$$

If now y is the semi-chord of the arc of circle-limit $2r$, we prove (§ 30) that $\dfrac{r}{\tan\ z} =$ constant.

Representing this constant by i, and making y tend toward zero, we have

$$\frac{2r}{2y} \doteq 1, \quad \text{whence}$$

$$2y \doteq 2\ i\ \tan\ z \doteq i\ \frac{I^{\frac{2y}{i}} - 1}{I^{\frac{y}{i}}},$$

or putting $\frac{2y}{i}=k$, $I=el$,

$$kI^{\frac{y}{i}}\dot{=}e^{kl}-1\dot{=}kl\;(1+\lambda),$$

λ being infinitesimal at the same time as k. Therefore, for the limit, $1=l$ and consequently $I=e$.

The circle traced on the sphere-limit with the arc r of the curve-limit for radius, has for length $2\pi r$. Therefore,

$$\odot y=2\pi r=2\pi i \tan z=\pi i\;(Y-Y^{-1}).$$

In the rectilineal $\triangle$ where a, β designate the angles opposite the sides a, b, we have (§ 25)

$$\sin a:\sin \beta=\odot a:\odot b=\pi i(A-A^{-1}):\pi i(B-B^{-1})$$
$$=\sin (a\sqrt{-1}):\sin (b\sqrt{-1}).$$

Thus in plane trigonometry as in spherical trigonometry, the sines of the angles are to each other as the sines of the opposite sides, only that on the sphere the sides are reals, and in the plane we must consider them as imaginaries, just as if the plane were an imaginary sphere.

We may arrive at this proposition without a preceding determination of the value of I.

If we designate the constant $\dfrac{r}{\tan z}$ by q, we shall have, as before

$$\odot y=\pi q\;(Y-Y^{-1}),$$

whence we deduce the same proportion as above, taking for i the distance for which the ratio I is equal to e.

If axiom XI is not true, there exists a determinate i, which must be substituted in the formulas.

If, on the contrary, this axiom is true, we must make in the formulas $i = \infty$. Because, in this case, the quantity $\frac{a}{\gamma} = Y$ is always $=1$, the sphere-limit being a plane, and the axes being parallel in Euclid's sense.

The exponent $\frac{y}{i}$ must therefore be zero, and consequently $i = \infty$.

It is easy to see that Bolyai's formulas of plane trigonometry are in accord with those of Lobachevski.

Take for example the formula of § 37,

$$\tan \text{ // }(a) = \sin B \tan \text{ // }(p),$$

a being the hypothenuse of a right-angled triangle, p one side of the right angle, and B the angle opposite to this side.

Bolyai's formula of § 31, I, gives

$$1 : \sin B = (A - A^{-1}) : (P - P^{-1}).$$

Now, putting for brevity, $\frac{1}{2}\text{// }(k) = k'$, we have $\tan 2p' : \tan 2a' = (\cot a' - \tan a') : (\cot p' - \tan p') = (A - A^{-1}) : (P - P^{-1}) = 1 : \sin B$.

APPENDIX III.

LIGHT FROM NON-EUCLIDEAN SPACES ON THE
TEACHING OF ELEMENTARY GEOMETRY.

By G. B. Halsted.

As foreshadowed by Bolyai and Riemann, founded by Cayley, extended and interpreted for hyperbolic, parabolic, elliptic spaces by Klein, recast and applied to mechanics by Sir Robert Ball, *projective metrics* may be looked upon as characteristic of what is highest and most peculiarly modern in all the bewildering range of mathematical achievement.

Mathematicians hold that number is wholly a creation of the human intellect, while on the contrary our space has an empirical element. Of possible geometries we can not say *a priori* which shall be that of our actual space, the space in which we move. Of course an advance so important, not only for mathematics but for philosophy, has had some metaphysical opponents, and as long ago as 1878 I mentioned in my Bibliography of Hyper-

Space and Non-Euclidean Geometry (American Journal of Mathematics, Vol. I, 1878, Vol. II, 1879) one of these, Schmitz-Dumont, as a sad paradoxer, and another, J. C. Becker, both of whom would ere this have shared the oblivion of still more antiquated fighters against the light, but that Dr. Schotten, praiseworthy for the very attempt at a comparative planimetry, happens to be himself a believer in the *a priori* founding of geometry, while his American reviewer, Mr. Ziwet, was then also an anti-non-Euclidean, though since converted.

He says, "we find that some of the best German text books do not try at all to define what is space, or what is a point, or even what is a straight line." Do any German geometries define space? I never remember to have met one that does.

In experience, what comes first is a bounded surface, with its boundaries, lines, and their boundaries, points. Are the points whose definitions are omitted anything different or better?

Dr. Schotten regards the two ideas "direction" and "distance" as intuitively given in the mind and as so simple as to not require definition.

When we read of two jockeys speeding

around a track in opposite directions, and also on page 87 of Richardson's Euclid, 1891, read, "The sides of the figure must be produced in the same direction of rotation; . . . going round the figure always in the same direction," we do not wonder that when Mr. Ziwet had written: "he therefore bases the definition of the straight line on these two ideas," he stops, modifies, and rubs that out as follows, "or rather recommends to elucidate the intuitive idea of the straight line possessed by any well-balanced mind by means of the still simpler ideas of direction" [in a circle] "and distance" [on a curve].

But when we come to geometry as a science, as foundation for work like that of Cayley and Ball, I think with Professor Chrystal: "It is essential to be careful with our definition of a *straight line*, for it will be found that virtually the properties of the straight line determine the nature of space.

"Our definition shall be that two points *in general* determine a straight line."

We presume that Mr. Ziwet glories in that unfortunate expression "a straight line is the shortest distance between two points," still occurring in Wentworth (New Plane Geometry, page 33), even after he has said, page 5, "the length of the straight line is called the *distance* between two points." If the *length* of the one straight line between two points is the distance between those points, how can the straight line itself be the *shortest* distance? If there is only one distance, it is the longest as much as the shortest distance, and if it is the *length* of this shorto-longest distance which is the distance then it is not the straight line itself which is the longo-shortest distance. But Wentworth also says: "Of all lines joining two points the *shortest* is the straight line."

This general comparison involves the measurement of curves, which involves the theory of limits, to say nothing of ratio. The very ascription of length to a curve involves the idea of a limit. And then to introduce this general axiom, as does Wentworth, only to prove a very special case of itself, that two sides of a triangle are together greater than the third, is surely bad logic, bad pedagogy, bad mathematics.

This latter theorem, according to the first of Pascal's rules for demonstrations, should not be proved at all, since every dog knows it. But to this objection, as old as the sophists, Simson long ago answered for the science of geometry, that the number of assumptions ought not to be increased without necessity ; or as Dedekind has it: " *Was beweisbar ist, soll in der Wissenschaft nicht ohne Beweis geglaubt werden.*"

Professor W. B. Smith (Ph. D., Goettingen), has written: " Nothing could be more unfortunate than the attempt to lay the notion of Direction at the bottom of Geometry."

Was it not this notion which led so good a mathematician as John Casey to give as a demonstration of a triangle's angle-sum the procedure called " a *practical* demonstration " on page 87 of Richardson's Euclid, and there described as "laying a 'straight edge' along one of the sides of the figure, and then turning it round so as to coincide with each side in turn."

This assumes that a segment of a straight line, a sect, may be translated without rotation, which assumption readily comes to view when you try the procedure in two-dimensional spherics. Though this fallacy was exposed by so eminent a geometer as Olaus Henrici in so public a place as the pages of 'Nature,' yet it has just been solemnly reproduced by Professor G. C. Edwards, of the University of California, in his Elements of Geometry: Mac-

Millan, 1895. It is of the greatest importance
for every teacher to know and connect the
commonest forms of assumption equivalent to
Euclid's Axiom XI. If in a plane two straight
lines perpendicular to a third nowhere meet,
are there others, not both perpendicular to
any third, which nowhere meet? Euclid's
Axiom XI is the assumption *No*. Playfair's
answers *no* more simply. But the very same
answer is given by the common assumption of
our geometries, usually unnoticed, that a circle
may be passed through any three points not
costraight.

This equivalence was pointed out by Bolyai
Farkas, who looks upon this as the simplest
form of the assumption. Other equivalents
are, the existence of any finite triangle whose
angle-sum is a straight angle; or the existence
of a plane rectangle; or that, in triangles, the
angle-sum is constant.

One of Legendre's forms was that through
every point within an angle a straight line
may be drawn which cuts both arms.

But Legendre never saw through this matter because he had not, as we have, the eyes
of Bolyai and Lobachevski to see with. The
same lack of their eyes has caused the author
of the charming book " Euclid and His Modern

Rivals," to give us one more equivalent form: "In any circle, the inscribed equilateral tetragon is greater than any one of the segments which lie outside it." (A New Theory of Parallels by C. L. Dodgson, 3d. Ed., 1890.)

Any attempt to define a straight line by means of "direction" is simply a case of "argumentum in circulo." In all such attempts the loose word "direction" is used in a sense which presupposes the straight line. The directions from a point in Euclidean space are only the ∞^2 rays from that point.

Rays not costraight can be said to have the same direction only after a theory of parallels is presupposed, assumed.

Three of the exposures of Professor G. C. Edwards' fallacy are here reproduced. The first, already referred to, is from *Nature,* Vol. XXIX, p. 453, March 13, 1884.

"I select for discussion the 'quaternion proof" given by Sir William Hamilton. . . . Hamilton's proof consists in the following:

"One side AB of the triangle ABC is turned about the point B till it lies in the continuation of BC; next, the line BC is made to slide along BC till B comes to C, and is then turned about C till it comes to lie in the continuation of AC.

"It is now again made to slide along CA till the point B comes to A, and is turned about A till it lies in the line AB. Hence it follows, *since rotation is independent of translation,* that the line has performed a whole revolution, that is, it has been turned through four right angles. But it has also described in succession the three exterior angles of the triangle, hence these are together equal to four right angles, and from this follows at once that the interior angles are equal to two right angles.

"To show how erroneous this reasoning is— in spite of Sir William Hamilton and in spite of quaternions—I need only point out that it holds exactly in the same manner for a triangle on the surface of the sphere, from which it would follow that the sum of the angles in a spherical triangle equals two right angles, whilst this sum is known to be always greater than two right angles. The proof depends only on the fact, that any line can be made to coincide with any other line, that two lines do so coincide when they have two points in common, and further, that a line may be turned about any point in it without leaving the surface. But if instead of the plane we take a spherical surface, and instead of a line a great circle on the sphere, all these conditions are again satisfied.

"The reasoning employed must therefore be fallacious, and the error lies in the words printed in italics; for these words contain an assumption which has not been proved.

"O. HENRICI."

Perronet Thompson, of Queen's College, Cambridge, in a book of which the third edition is dated 1830, says:

"Professor Playfair, in the Notes to his 'Elements of Geometry' [1813], has proposed another demonstration, founded on a remarkable *non causa pro causa*.

"It purports to collect the fact [Eu. I., 32, Cor., 2] that (on the sides being successively prolonged to the same hand) the exterior angles of a rectilinear triangle are together equal to four right angles, from the circumstance that a straight line carried round the perimeter of a triangle by being applied to all the sides in succession, is brought into its old situation again; the argument being, that because this line has made the sort of somerset it would do by being turned through four right angles about a fixed point, the exterior angles of the triangle have necessarily been equal to four right angles.

"The answer to which is, that there is no connexion between the things at all, and that the result will just as much take place where the exterior angles are avowedly not equal to four right angles.

"Take, for example, the plane triangle formed by three small arcs of the same or equal circles,

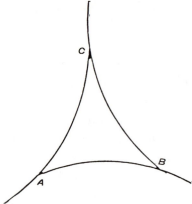

as in the margin; and it is manifest that an arc of this circle may be carried round precisely in the way described and return to its old situation, and yet there be no pretense for inferring that the exterior angles were equal to four right angles.

"And if it is urged that these are *curved* lines and the statement made was of straight; then the answer is by demanding to know, what property of straight lines has been laid down or established, which determines that what is not true in the case of other lines is true in theirs. It has been shown that, as a general proposition, the connexion between a line returning to its place and the exterior angles having been equal to four right angles, is a *non sequitur;* that it is a thing that may be or may not be; that the notion that it returns to its place *because* the exterior angles have been equal to four right angles, is a mistake. From which it is a legitimate conclusion, that if it had pleased nature to make the exterior angles of a triangle greater or less than four right angles, this would not have created the smallest impediment to the line's returning to its old situation after being carried round the sides; and consequently the line's returning is no evidence of the angles not being greater or less than four right angles."

Charles L. Dodgson, of Christ Church, Oxford, in his "Curiosa Mathematica," Part I, pp. 70–71, 3d Ed., 1890, says:

"Yet another process has been invented— quite fascinating in its brevity and its elegance—which, though involving the same fallacy as the Direction-Theory, proves Euc. I, 32, without even mentioning the dangerous word 'Direction.'

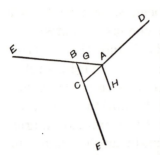

"We are told to take any triangle A B C; to produce C A to D; to make part of CD, viz., A D, revolve, about A, into the position A B E; then to make part of this line, viz., BE, revolve, about B, into the position BCF; and lastly to make part of this line, viz., CF, revolve, about C, till it lies along CD, of which it originally formed a part. We are then assured that it must have revolved through four right angles: from which it easily follows that the interior angles of the triangle are together equal to two right angles.

"The disproof of this fallacy is almost as brief and elegant as the fallacy itself. We first quote the general principle that we can not reasonably be told to make a line fulfill *two* conditions, either of which is enough by itself to fix its position: e. g., given three points X, Y, Z, we can not reasonably be told to draw a line from X which shall pass through Y *and* Z: we can make it pass through Y, but it must then take its chance of passing through Z; and *vice versa*.

"Now let us suppose that, while one part of

AE, viz., BE, revolves into the position BF, another little bit of it, viz., AG, revolves, through an equal angle, into the position AH; and that, while CF revolves into the position of lying along CD, AH revolves—and here comes the fallacy.

"You must not say 'revolves, through an equal angle, into the position of lying along AD,' for this would be to make AH *fulfill two conditions at once.*

"If you say that the one condition involves the other, you are virtually asserting that the lines CF, AH are equally inclined to CD—and this in *consequence* of AH having been so drawn that these same lines are equally inclined to AE.

"That is, you are asserting, 'A pair of lines which are equally inclined to a certain transversal, are so to any transversal.' [Deducible from Euc. I, 27, 28, 29.]"

GEOMETRICAL RESEARCHES

ON THE THEORY OF

PARALLELS

BY

NICHOLAS LOBACHEVSKI

TRANSLATED BY

DR. GEORGE BRUCE HALSTED

TRANSLATOR'S PREFACE.

Lobachevski was the first man ever to publish a non-Euclidean geometry.

Of the immortal essay now first appearing in English Gauss said, "The author has treated the matter with a master-hand and in the true geometer's spirit. I think I ought to call your attention to this book, whose perusal can not fail to give you the most vivid pleasure."

Clifford says, "It is quite simple, merely Euclid without the vicious assumption, but the way things come out of one another is quite lovely." * * * "What Vesalius was to Galen, what Copernicus was to Ptolemy, that was Lobachevski to Euclid."

Says Sylvester, "In Quaternions the example has been given of Algebra released from the yoke of the commutative principle of multiplication—an emancipation somewhat akin to Lobachevski's of Geometry from Euclid's noted empirical axiom."

Cayley says, "It is well known that Euclid's twelfth axiom, even in Playfair's form of it, has been considered as needing demonstration; and that Lobachevski constructed a perfectly consistent theory, wherein this axiom was assumed not to hold good, or say a system of non-Euclidean plane geometry. There is a like system of non-Euclidean solid geometry."

GEORGE BRUCE HALSTED.

2407 San Marcos Street,
    Austin, Texas.
May 1, 1891.

TRANSLATOR'S INTRODUCTION.

"Prove all things, hold fast that which is good," does not mean demonstrate everything. From nothing assumed, nothing can be proved. "Geometry without axioms," was a book which went through several editions, and still has historical value. But now a volume with such a title would, without opening it, be set down as simply the work of a paradoxer.

The set of axioms far the most influential in the intellectual history of the world was put together in Egypt; but really it owed nothing to the Egyptian race, drew nothing from the boasted lore of Egypt's priests.

The Papyrus of the Rhind, belonging to the British Museum, but given to the world by the erudition of a German Egyptologist, Eisenlohr, and a German historian of mathematics, Cantor, gives us more knowledge of the state of mathematics in ancient Egypt than all else previously accessible to the modern world. Its whole testimony confirms with overwhelming force the position that Geometry as a science, strict and self-conscious deductive reasoning, was created by the subtle intellect of the same race whose bloom in art still overawes us in the Venus of Milo, the Apollo Belvidere, the Laocoon.

In a geometry occur the most noted set of axioms, the geometry of Euclid, a pure Greek, professor at the University of Alexandria.

Not only at its very birth did this typical product of the Greek genius assume sway as ruler in the pure sciences, not only does its first efflorescence carry us through the splendid days of Theon and Hypatia, but unlike the latter, fanatics can not murder it; that dismal flood, the dark ages, can not drown it. Like the phœnix of its native Egypt, it rises with the new birth of culture. An Anglo-Saxon, Adelard of Bath, finds it clothed in Arabic vestments in the land of the Alhambra. Then clothed in Latin, it and the new-born printing press confer honor on each other. Finally back again in its original Greek, it is published first in queenly Basel, then in stately Oxford. The latest edition in Greek is from Leipsic's learned presses.

How the first translation into our cut-and-thrust, survival-of-the-fittest English was made from the Greek and Latin by Henricus Billingsly, Lord Mayor of London, and published with a preface by John Dee the Magician, may be studied in the Library of our own Princeton, where they have, by some strange chance, Billingsly's own copy of the Arabic-Latin version of Campanus bound with the Editio Princeps in Greek and enriched with his autograph emendations. Even to-day in the vast system of examinations set by Cambridge, Oxford, and the British government, no proof will be accepted which infringes Euclid's order, a sequence founded upon his set of axioms.

The American ideal is success. In twenty years the American maker expects to be improved upon, superseded. The Greek ideal was perfection. The Greek Epic and Lyric poets, the Greek sculptors, remain unmatched. The axioms of the Greek geometer remained unquestioned for twenty centuries.

How and where doubt came to look toward them is of no ordinary interest, for this doubt was epoch-making in the history of mind.

Among Euclid's axioms was one differing from the others in prolixity, whose place fluctuates in the manuscripts, and which is not used in Euclid's first twenty-seven propositions. Moreover it is only then brought in to prove the inverse of one of these already demonstrated.

All this suggested, at Europe's renaissance, not a doubt of the axiom, but the possibility of getting along without it, of deducing it from the other axioms and the twenty-seven propositions already proved. Euclid demonstrates things more axiomatic by far. He proves what every dog knows, that any two sides of a triangle are together greater than the third. Yet when he has perfectly proved that lines making with a transversal equal alternate angles are parallel, in order to prove the inverse, that parallels cut by a transversal make equal alternate angles, he brings in the unwieldly postulate or axiom:

"If a straight line meet two straight lines, so as to make the two interior angles on the same side of it taken together less than two right angles, these straight lines, being continually produced, shall at length meet on that side on which are the angles which are less than two right angles."

Do you wonder that succeeding geometers wished by demonstration to push this unwieldly thing from the set of fundamental axioms.

Numerous and desperate were the attempts to deduce it from reasonings about the nature of the straight line and plane angle. In the "Encyclopœdie der Wissenschaften und Kunste; Von Ersch und Gruber;" Leipzig, 1838; under "Parallel," Sohncke says that in mathematics there is nothing over which so much has been spoken, written, and striven, as over the theory of parallels, and all, so far (up to his time), without reaching a definite result and decision.

Some acknowledged defeat by taking a new definition of parallels, as for example the stupid one, "Parallel lines are everywhere equally distant," still given on page 33 of Schuyler's Geometry, which that author, like many of his unfortunate prototypes, then attempts to identify with Euclid's definition by pseudo-reasoning which tacitly assumes Euclid's postulate, e. g. he says p. 35: "For, if not parallel, they are not everywhere equally distant; and since they lie in the same plane; must approach when produced one way or the other; and since straight lines continue in the same direction, must continue to approach if produced farther, and if sufficiently produced, must meet." This is nothing but Euclid's assumption, diseased and contaminated by the introduction of the indefinite term "direction."

How much better to have followed the third class of his predecessors who honestly assume a new axiom differing from Euclid's in form if not in essence. Of these the best is that called Playfair's; "Two lines which intersect can not both be parallel to the same line."

The German article mentioned is followed by a carefully prepared list of ninety-two authors on the subject. In English an account of like attempts was given by Perronet Thompson, Cambridge, 1833, and is brought up to date in the charming volume, "Euclid and his Modern Rivals," by C. L. Dodgson, late Mathematical Lecturer of Christ Church, Oxford, the Lewis Carroll, author of Alice in Wonderland.

All this shows how ready the world was for the extraordinary flaming-forth of genius from different parts of the world which was at once to overturn, explain, and remake not only all this subject but as consequence all philosophy, all ken-lore. As was the case with the discovery of the Conservation of .Energy, the independent irruptions of genius, whether in Russia, Hungary, Germany, or even in Canada gave everywhere the same results.

At first these results were not fully understood even by the brightest intellects. Thirty years after the publication of the book he mentions, we see the brilliant Clifford writing from Trinity College, Cambridge, April 2, 1870, "Several new ideas have come to me lately: First I have procured Lobachevski, 'Études Géométriques sur la Théorie des Parallels' - - - a small tract of which Gauss, therein quoted, says: L'auteur a traité la matière en main de maître et avec le véritable esprit géométrique. Je crois devoir appeler votre attention sur ce livre, dont la lecture ne peut manquer de vous causer le plus vif plaisir.'" Then says Clifford: "It is quite simple, merely Euclid without the vicious assumption, but the way the things come out of one another is quite lovely."

The first axiom doubted is called a "vicious assumption," soon no man sees more clearly than Clifford that all are assumptions and none vicious. He had been reading the French translation by Hoüel, published in 1866, of a little book of 61 pages published in 1840 in Berlin under the title Geometrische Untersuchungen zur Theorie der Parallellinien by Nicolas Lobachevski (1793-1856), the first public expression of whose discoveries, however, dates back to a discourse at Kasan on February 12, 1826.

Under this commonplace title who would have suspected the discovery of a new space in which to hold our universe and ourselves.

A new kind of universal space; the idea is a hard one. To name it, all the space in which we think the world and stars live and move and have their being was ceded to Euclid as his by right of pre-emption, description, and occupancy; then the new space and its quick-following fellows could be called Non-Euclidean.

Gauss in a letter to Schumacher, dated Nov. 28, 1846, mentions that as far back as 1792 he had started on this path to a new universe. Again he says: "La géométrie non-euclidienne ne renferme en elle rien de contradictoire, quoique, à prèmiere vue, beaucoup de ses résultats aien l'air de paradoxes. Ces contradictions apparents doivent être regardées comme l'effet d'une illusion, due à l'habitude que nous avons prise de bonne heure de considérer la géométrie euclidienne comme rigoureuse."

But here we see in the last word the same imperfection of view as in Clifford's letter. The perception has not yet come that though the non-Euclidean geometry is rigorous, Euclid is not one whit less so.

A former friend of Gauss at Gœttingen was the Hungarian Wolfgang Bolyai. His principal work, published by subscription, has the following title:

Tentamen Juventutem studiosam in elementa Matheseos purae, elementaris ac sublimioris, methodo intuitiva, evidentiaque huic propria, introducendi. Tomus Primus, 1832; Secundus, 1833. 8vo. Maros-Vásárhelyini.

In the first volume with special numbering, appeared the celebrated Appendix of his son John Bolyai with the following title:

APPENDIX.

SCIENTIAM SPATII *absolute veram* exhibens: *a veritate aut falsitate Axiomatis XI Euclidei (a priori haud unquam decidenda) independentem.* Auctore JOHANNE BOLYAI de eadem, Geometrarum in Exercitu Caesareo Regio Austriaco Castrensium Capitaneo. (26 pages of text).

This marvellous Appendix has been translated into French, Italian, English and German.

In the title of Wolfgang Bolyai's last work, the only one he composed in German (88 pages of text, 1851), occurs the following:

"und da die Frage, *ob zwey von der dritten geschnittene Geraden, wenn die summe der inneren Winkel nicht=2R, sich schneiden oder nicht?* niemand auf der Erde ohne ein Axiom (wie *Euclid* das XI) aufzustellen, beantworten wird; die davon unabhængige Geometrie abzusondern; und eine auf die *Ja*-Antwort, andere auf das *Nein* so zu bauen, dass die Formeln der letzen, auf ein Wink auch in der ersten gültig seyen."

The author mentions Lobachevski's Geometrische Untersuchungen, Berlin, 1840, and compares it with the work of his son John Bolyai, "au sujet duquel il dit: 'Quelques exemplaires de l'ouvrage publié ici ont été envoyés à cette époque à Vienne, à Berlin, à Gœttingue. . . De Goettingue le géant mathématique, [Gauss] qui du sommet des hauteurs embrasse du même regard les astres et la profondeur des abîmes, a écrit qu'il était ravi de voir exécuté le travail qu'il avait commencé pour le laisser après lui dans ses papiers.' "

In fact this first of the Non-Euclidean geometries accepts all of Euclid's axioms but the last, which it flatly denies and replaces by its contradictory, that the sum of the interior angles made on the same side of a transversal by two straight lines may be less than a straight angle without the lines meeting. A perfectly consistent and elegant geometry then follows, in which the sum of the angles of a triangle is always less than a straight angle, and not every triangle has its vertices concyclic.

THEORY OF PARALLELS.

In geometry I find certain imperfections which I hold to be the reason why this science, apart from transition into analytics, can as yet make no advance from that state in which it has come to us from Euclid.

As belonging to these imperfections, I consider the obscurity in the fundamental concepts of the geometrical magnitudes and in the manner and method of representing the measuring of these magnitudes, and finally the momentous gap in the theory of parallels, to fill which all efforts of mathematicians have been so far in vain.

For this theory Legendre's endeavors have done nothing, since he was forced to leave the only rigid way to turn into a side path and take refuge in auxiliary theorems which he illogically strove to exhibit as necessary axioms. My first essay on the foundations of geometry I published in the Kasan *Messenger* for the year 1829. In the hope of having satisfied all requirements, I undertook hereupon a treatment of the whole of this science, and published my work in separate parts in the "*Gelehrten Schriften der Universitæt Kasan*" for the years 1836, 1837, 1838, under the title "New Elements of Geometry, with a complete Theory of Parallels." The extent of this work perhaps hindered my countrymen from following such a subject, which since Legendre had lost its interest. Yet I am of the opinion that the Theory of Parallels should not lose its claim to the attention of geometers, and therefore I aim to give here the substance of my investigations, remarking beforehand that contrary to the opinion of Legendre, all other imperfections—for example, the definition of a straight line—show themselves foreign here and without any real influence on the theory of parallels.

In order not to fatigue my reader with the multitude of those theorems whose proofs present no difficulties, I prefix here only those of which a knowledge is necessary for what follows.

1. A straight line fits upon itself in all its positions. By this I mean that during the revolution of the surface containing it the straight line does not change its place if it goes through two unmoving points in the surface: (*i. e.*, if we turn the surface containing it about two points of the line, the line does not move.)

2. Two straight lines can not intersect in two points.

3. A straight line sufficiently produced both ways must go out beyond all bounds, and in such way cuts a bounded plain into two parts.

4. Two straight lines perpendicular to a third never intersect, how far soever they be produced.

5. A straight line always cuts another in going from one side of it over to the other side: (*i. e.*, one straight line must cut another if it has points on both sides of it.)

6. Vertical angles, where the sides of one are productions of the sides of the other, are equal. This holds of plane rectilineal angles among themselves, as also of plane surface angles: (*i. e.*, dihedral angles.)

7. Two straight lines can not intersect, if a third cuts them at the same angle.

8. In a rectilineal triangle equal sides lie opposite equal angles, and inversely.

9. In a rectilineal triangle, a greater side lies opposite a greater angle. In a right-angled triangle the hypothenuse is greater than either of the other sides, and the two angles adjacent to it are acute.

10. Rectilineal triangles are congruent if they have a side and two angles equal, or two sides and the included angle equal, or two sides and the angle opposite the greater equal, or three sides equal.

11. A straight line which stands at right angles upon two other straight lines not in one plane with it is perpendicular to all straight lines drawn through the common intersection point in the plane of those two.

12. The intersection of a sphere with a plane is a circle.

13. A straight line at right angles to the intersection of two perpendicular planes, and in one, is perpendicular to the other.

14. In a spherical triangle equal sides lie opposite equal angles, and inversely.

15. Spherical triangles are congruent (or symmetrical) if they have two sides and the included angle equal, or a side and the adjacent angles equal.

From here follow the other theorems with their explanations and proofs.

16. All straight lines which in a plane go out from a point can, with reference to a given straight line in the same plane, be divided into two classes — into *cutting* and *not-cutting*.

The *boundary lines* of the one and the other class of those lines will be called *parallel to the given line*.

From the point A (Fig. 1) let fall upon the line BC the perpendicular AD, to which again draw the perpendicular AE.

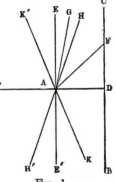

In the right angle EAD either will all straight lines which go out from the point A meet the line DC, as for example AF, or some of them, like the perpendicular AE, will not meet the line DC.   In the uncertainty whether the perpendicular AE is the only line which does not meet DC, we will assume it may be possible that there are still other lines, for example AG,

Fig. 1.

which do not cut DC, how far soever they may be prolonged.   In passing over from the cutting lines, as AF, to the not-cutting lines, as AG, we must come upon a line AH, parallel to DC, a boundary line, upon one side of which all lines AG are such as do not meet the line DC, while upon the other side every straight line AF cuts the line DC.

The angle HAD between the parallel HA and the perpendicular AD is called the parallel angle (angle of parallelism), which we will here designate by $\varPi$ (p) for AD = p.

If $\varPi$ (p) is a right angle, so will the prolongation AE′ of the perpendicular AE likewise be parallel to the prolongation DB of the line DC, in addition to which we remark that in regard to the four right angles, which are made at the point A by the perpendiculars AE and AD, and their prolongations AE′ and AD′, every straight line which goes out from the point A, either itself or at least its prolongation, lies in one of the two right angles which are turned toward BC, so that except the parallel EE′ all others, if they are sufficiently produced both ways, must intersect the line BC.

If $\varPi$ (p) $< \frac{1}{2}$ π, then upon the other side of AD, making the same angle DAK = $\varPi$ (p) will lie also a line AK, parallel to the prolongation DB of the line DC, so that under this assumption we must also make a distinction of *sides in parallelism*.

All remaining lines or their prolongations within the two right angles turned toward BC pertain to those that intersect, if they lie within the angle HAK = 2 $\varPi$ (p) between the parallels; they pertain on the other hand to the non-intersecting AG, if they lie upon the other sides of the parallels AH and AK, in the opening of the two angles EAH = $\frac{1}{2}\pi$ — $\varPi$ (p), E'AK = $\frac{1}{2}\pi$ — $\varPi$ (p), between the parallels and EE' the perpendicular to AD. Upon the other side of the perpendicular EE' will in like manner the prolongations AH' and AK' of the parallels AH and AK likewise be parallel to BC; the remaining lines pertain, if in the angle K'AH', to the intersecting, but if in the angles K'AE, H'AE' to the non-intersecting.

In accordance with this, for the assumption $\varPi$ (p) = $\frac{1}{2}\pi$. the lines can be only intersecting or parallel; but if we assume that $\varPi$ (p) < $\frac{1}{2}\pi$, then we must allow two parallels, one on the one and one on the other side; in addition we must distinguish the remaining lines into non-intersecting and intersecting.

For both assumptions it serves as the mark of parallelism that the line becomes intersecting for the smallest deviation toward the side where lies the parallel, so that if AH is parallel to DC, every line AF cuts DC, how small soever the angle HAF may be.

17. *A straight line maintains the characteristic of parallelism at all its points.*

Given AB (Fig. 2) parallel to CD, to which latter AC is **perpendic**

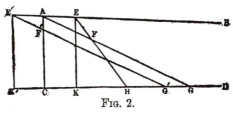

FIG. 2.

ular. We will consider two points taken at random on the line AB **and** its production beyond the perpendicular.

Let the point E lie on that side of the perpendicular on which AB is looked upon as parallel to CD.

Let fall from the point E a perpendicular EK on CD and so draw EF that it falls within the angle BEK.

Connect the points A and F by a straight line, whose production then (by Theorem 16) must cut CD somewhere in G. Thus we get a triangle ACG, into which the line EF goes; now since this latter, from the construction, can not cut AC, and can not cut AG or EK a second time (Theorem 2), therefore it must meet CD somewhere at H (Theorem 3).

Now let E' be a point on the production of AB and E'K' perpendicular to the production of the line CD; draw the line E'F' making so small an angle AE'F' that it cuts AC somewhere in F'; making the same angle with AB, draw also from A the line AF, whose production will cut CD in G (Theorem 16.)

Thus we get a triangle AGC, into which goes the production of the line E'F'; since now this line can not cut AC a second time, and also can not cut AG, since the angle BAG = BE'G', (Theorem 7), therefore must it meet CD somewhere in G'.

Therefore from whatever points E and E' the lines EF and E'F' go out, and however little they may diverge from the line AB, yet will they always cut CD, to which AB is parallel.

18. *Two lines are always mutually parallel.*

Let AC be a perpendicular on CD, to which AB is parallel if we draw from C the line CE making any acute angle ECD with CD, and let fall from A the perpendicular AF upon CE, we obtain a right-angled triangle ACF, in which AC, being the hypothenuse, is greater than the side AF (Theorem 9.)

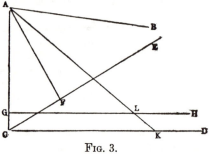

Make AG = AF, and slide

Fig. 3.

the figure EFAB until AF coincides with AG, when AB and FE will take the position AK and GH, such that the angle BAK = FAC, consequently AK must cut the line DC somewhere in K (Theorem 16), thus forming a triangle AKC, on one side of which the perpendicular GH intersects the line AK in L (Theorem 3), and thus determines the dis- tance AL of the intersection point of the lines AB and CE on the line AB from the point A.

Hence it follows that CE will always intersect AB, how small soever may be the angle ECD, consequently CD is parallel to AB (Theorem 16.)

19. *In a rectilineal triangle the sum of the three angles can not be greater than two right angles.*

Suppose in the triangle ABC (Fig. 4) the sum of the three angles is equal to $\pi + a$; then choose in case of the inequality of the sides the smallest BC, halve it in D, draw from A through D the line AD and make the prolongation of it, DE, equal to AD, then join the point E to the point C by the

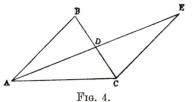

Fig. 4.

straight line EC. In the congruent triangles ADB and CDE, the angle ABD = DCE, and BAD = DEC (Theorems 6 and 10); whence follows that also in the triangle ACE the sum of the three angles must be equal to $\pi + a$; but also the smallest angle BAC (Theorem 9) of the triangle ABC in passing over into the new triangle ACE has been cut up into the two parts EAC and AEC. Continuing this process, continually halving the side opposite the smallest angle, we must finally attain to a triangle in which the sum of the three angles is $\pi + a$, but wherein are two angles, each of which in absolute magnitude is less than $\frac{1}{2}a$; since now, however, the third angle can not be greater than π, so must a be either null or negative.

20. *If in any rectilineal triangle the sum of the three angles is equal to two right angles, so is this also the case for every other triangle.*

If in the rectilineal triangle ABC (Fig. 5) the sum of the three angles $= \pi$, then must at least two of its angles, A and C, be acute. Let fall from the vertex of the third angle B upon the opposite side AC the perpendicular p. This will cut the tri-angle into two right-angled triangles, in each

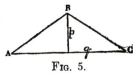

Fɪɢ. 5.

of which the sum of the three angles must also be π, since it can not in either be greater than π, and in their combination not less than π.

So we obtain a right-angled triangle with the perpendicular sides p and q, and from this a quadrilateral whose opposite sides are equal and whose adjacent sides p and q are at right angles (Fig. 6.)

By repetition of this quadrilateral we can make another with sides np and q, and finally a quadrilateral ABCD with sides at right angles to each other, such that AB = np, AD = mq, DC = np, BC = mq, where

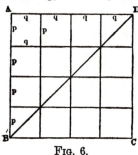

Fɪɢ. 6.

m and n are any whole numbers. Such a quadrilateral is divided by the diagonal DB into two congruent right-angled triangles, BAD and BCD, in each of which the sum of the three angles $= \pi$.

The numbers n and m can be taken sufficiently great for the right-angled triangle ABC (Fig. 7) whose perpendicular sides AB = np, BC = mq, to enclose within itself another given (right-angled) triangle BDE as soon as the right-angles fit each other.

2 — par.

Drawing the line DC, we obtain right-angled triangles of which every successive two have a side in common.

The triangle ABC is formed by the union of the two triangles ACD and DCB, in neither of which can the sum of the angles be greater than π; consequently it must be equal to π, in order that the sum in the compound triangle may be equal to π.

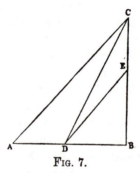

Fig. 7.

In the same way the triangle BDC consists of the two triangles DEC and DBE, consequently must in DBE the sum of the three angles be equal to π, and in general this must be true for every triangle, since each can be cut into two right-angled triangles.

From this it follows that only two hypotheses are allowable: Either is the sum of the three angles in all rectilineal triangles equal to π, or this sum is in all less than π.

21. *From a given point we can always draw a straight line that shall make with a given straight line an angle as small as we choose.*

Let fall from the given point A (Fig. 8) upon the given line BC the

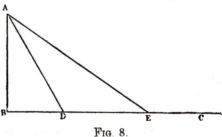

Fig. 8.

perpendicular AB; take upon BC at random the point D; draw the line AD; make DE = AD, and draw AE.

In the right-angled triangle ABD let the angle ADB = a; then must in the isosceles triangle ADE the angle AED be either $\frac{1}{2}a$ or less (Theorems 8 and 20). Continuing thus we finally attain to such an angle, AEB, as is less than any given angle.

22. *If two perpendiculars to the same straight line are parallel to each other, then the sum of the three angles in a rectilineal triangle is equal to two right angles.*

Let the lines AB and CD (Fig. 9) be parallel to each other and perpendicular to AC.

Draw from A the lines AE and AF to the points E and F, which are taken on the line CD at any distances FC > EC from the point C.

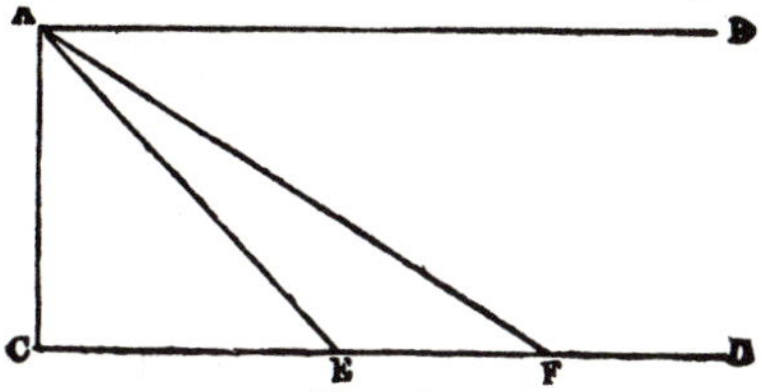

Suppose in the right-angled triangle ACE the sum of the three angles is equal to $\pi - a$, in the triangle AEF equal to $\pi - \beta$, then must it in triangle ACF equal $\pi - a - \beta$, where a and β can not be negative.

FIG. 9.

Further, let the angle BAF = a, AFC = b, so is $a + \beta$ = a — b; now by revolving the line AF away from the perpendicular AC we can make the angle a between AF and the parallel AB as small as we choose; so also can we lessen the angle b, consequently the two angles a and β can have no other magnitude than $a = 0$ and $\beta = 0$.

It follows that in all rectilineal triangles the sum of the three angles is either π and at the same time also the parallel angle Π (p) = $\frac{1}{2}$ π for every line p, or for all triangles this sum is $< \pi$ and at the same time also Π(p) $< \frac{1}{2}$ π.

The first assumption serves as *foundation for the ordinary geometry and plane trigonometry.*

The second assumption can likewise be admitted without leading to any contradiction in the results, and founds a new geometric science, to which I have given the name *Imaginary Geometry,* and which I intend here to expound as far as the development of the equations between the sides and angles of the rectilineal and spherical triangle.

23. *For every given angle a there is a line p such that Π (p) = a.*

Let AB and AC (Fig. 10) be two straight lines which at the intersection point A make the acute angle a; take at random on AB a point

B'; from this point drop B'A' at right angles to AC; make A'A″ =
AA'; erect at A″ the perpendicular A″B″; and so continue until a per-

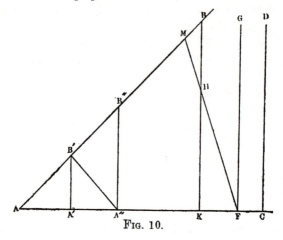

Fig. 10.

pendicular CD is attained, which no longer intersects AB. This must
of necessity happen, for if in the triangle AA'B' the sum of all three
angles is equal to π — a, then in the triangle AB'A″ it equals π — 2a,
in triangle AA″B″ less than π — 2a (Theorem 20), and so forth, until
it finally becomes negative and thereby shows the impossibility of constructing the triangle.

The perpendicular CD may be the very one nearer than which to the
point A all others cut AB; at least in the passing over from those that
cut to those not cutting such a perpendicular FG must exist.

Draw now from the point F the line FH, which makes with FG the
acute angle HFG, on that side where lies the point A. From any point
H of the line FH let fall upon AC the perpendicular HK, whose prolongation consequently must cut AB somewhere in B, and so makes a
triangle AKB, into which the prolongation of the line FH enters, and
therefore must meet the hypothenuse AB somewhere in M. Since the
angle GFH is arbitrary and can be taken as small as we wish, therefore
FG is parallel to AB and AF = p. (Theorems 16 and 18.)

One easily sees that with the lessening of p the angle α increases, while,
for p = 0, it approaches the value $\frac{1}{2}\pi$; with the growth of p the angle
α decreases, while it continually approaches zero for p = ∞ .

Since we are wholly at liberty to choose what angle we will under stand by the symbol Π (p) when the line p is expressed by a negative number, so we will assume

$$\Pi(\mathrm{p}) + \Pi(-\mathrm{p}) = \pi,$$

an equation which shall hold for all values of p, positive as well as negative, and for p $= 0$.

24. *The farther parallel lines are prolonged on the side of their parallelism, the more they approach one another.*

If to the line AB (Fig. 11) two perpendiculars AC $=$ BD are erected and their end-points C and D joined by a straight line, then will the quadrilateral CABD have two right angles at A and B, but two acute angles at C and D (Theorem 22) which are equal to one another, as we can easily see by thinking the quadrilateral super-

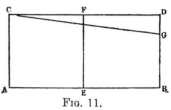

Fɪɢ. 11.

imposed upon itself so that the line BD falls upon AC and AC upon BD.

Halve AB and erect at the mid-point E the line EF perpendicular to AB. This line must also be perpendicular to CE, since the quadrilaterals CAEF and FDBE fit one another if we so place one on the other that the line EF remains in the same position. Hence the line CD can not be parallel to AB, but the parallel to AB for the point C, namely CG, must incline toward AB (Theorem 16) and cut from the perpendicular BD a part BG $<$ CA.

Since C is a random point in the line CG, it follows that CG itself nears AB the more the farther it is prolonged.

25. *Two straight lines which are parallel to a third are also parallel to each other.*

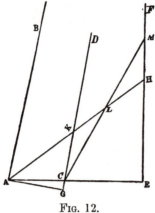

Fig. 12.

We will first assume that the three lines AB, CD, EF (Fig. 12) lie in one plane. If two of them in order, AB and CD, are parallel to the outmost one, EF, so are AB and CD parallel to each other. In order to prove this, let fall from any point A of the outer line AB upon the other outer line FE, the perpendicular AE, which will cut the middle line CD in some point C (Theorem 3), at an angle $DCE < \frac{1}{2}\pi$ on the side toward EF, the parallel to CD (Theorem 22).

A perpendicular AG let fall upon CD from the same point, A, must fall within the opening of the acute angle ACG (Theorem 9); every other line AH from A drawn within the angle BAC must cut EF, the parallel to AB, somewhere in H, how small soever the angle BAH may be; consequently will CD in the triangle AEH cut the line AH somewhere in K, since it is impossible that it should meet EF. If AH from the point A went out within the angle CAG, then must it cut the prolongation of CD between the points C and G in the triangle CAG. Hence follows that AB and CD are parallel (Theorems 16 and 18).

Were both the outer lines AB and EF assumed parallel to the middle line CD, so would every line AK from the point A, drawn within the angle BAE, cut the line CD somewhere in the point K, how small soever the angle BAK might be.

Upon the prolongation of AK take at random a point L and join it with C by the line CL, which must cut EF somewhere in M, thus making a triangle MCE.

The prolongation of the line AL within the triangle MCE can cut neither AC nor CM a second time, consequently it must meet EF somewhere in H; therefore AB and EF are mutually parallel.

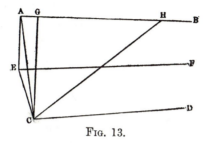

Fig. 13.

Now let the parallels AB and CD (Fig. 13) lie in two planes whose intersection line is EF. From a random point E of this latter let fall a perpendicular EA upon one of the two parallels, *e. g.*, upon AB, then from A, the foot of the perpendicular EA, let fall a new perpendicular AC upon the other parallel CD and join the end-points E and C of the two perpendiculars by the line EC. The angle BAC must be acute (Theorem 22), consequently a perpendicular CG from C let fall upon AB meets it in the point G upon that side of CA on which the lines AB and CD are considered as parallel.

Every line EH [in the plane FEAB], however little it diverges from EF, pertains with the line EC to a plane which must cut the plane of the two parallels AB and CD along some line CH. This latter line cuts AB somewhere, and in fact in the very point H which is common to all three planes, through which necessarily also the line EH goes; consequently EF is parallel to AB.

In the same way we may show the parallelism of EF and CD.

Therefore the hypothesis that a line EF is parallel to one of two other parallels, AB and CD, is the same as considering EF as the intersection of two planes in which two parallels, AB, CD, lie.

Consequently two lines are parallel to one another if they are parallel to a third line, though the three be not co-planar.

The last theorem can be thus expressed:

Three planes intersect in lines which are all parallel to each other if the parallelism of two is pre-supposed.

26. *Triangles standing opposite to one another on the sphere are equivalent in surface.*

By opposite triangles we here understand such as are made on both sides of the center by the intersections of the sphere with planes; in such triangles, therefore, the sides and angles are in contrary order.

In the opposite triangles ABC and A′B′C′ (Fig. 14, where one of them must be looked upon as represented turned about), we have the sides AB = A′B′, BC = B′C′, CA = C′A′, and the corresponding angles

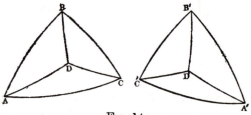

FIG. 14.

at the points A, B, C are likewise equal to those in the other triangle at th points A′, B′, C′.

Through the three points A, B, C, suppose a plane passed, and upon it from the center of the sphere a perpendicular dropped whose prolongations both ways cut both opposite triangles in the points D and D′ of the sphere. The distances of the first D from the points ABC, in arcs of great circles on the sphere, must be equal (Theorem 12) as well to each other as also to the distances D′A′, D′B′, D′C′, on the other triangle (Theorem 6), consequently the isosceles triangles about the points D and D′ in the two spherical triangles ABC and A′B′C′ are congruent.

In order to judge of the equivalence of any two surfaces in general, I take the following theorem as fundamental:

Two surfaces are equivalent when they arise from the mating or separating of equal parts.

27. *A three-sided solid angle equals the half sum of the surface angles less a right-angle.*

In the spherical triangle ABC (Fig. 15), where each side $< \pi$, designate the angles by A, B, C; prolong the side AB so that a whole circle ABA′B′A is produced; this divides the sphere into two equal parts.

In that half in which is the triangle ABC, prolong now the other two sides through their common intersection point C until they meet the circle in A′ and B′.

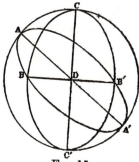

Fɪɢ. 15.

In this way the hemisphere is divided into four triangles, ABC, ACB′, B′CA′, A′CB, whose size may be designated by P, X, Y, Z. It is evident that here $P + X = B$, $P + Z = A$.

The size of the spherical triangle Y equals that of the opposite triangle ABC′, having a side AB in common with the triangle P, and whose third angle C′ lies at the end-point of the diameter of the sphere which goes from C through the center D of the sphere (Theorem 26). Hence it follows that

$P + Y = C$, and since $P + X + Y + Z = \pi$, therefore we have also
$$P = \tfrac{1}{2}(A + B + C - \pi).$$

We may attain to the same conclusion in another way, based solely upon the theorem about the equivalence of surfaces given above. (Theorem 26.)

In the spherical triangle ABC (Fig. 16), halve the sides AB and BC, and through the mid-points D and E draw a great circle; upon this let fall from A, B, C the perpendiculars AF, BH, and CG. If the perpendicular from B falls at H between D and E, then will of the triangles so made BDH = AFD, and BHE = EGC (Theorems 6 and 15), whence follows that

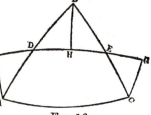

Fɪɢ. 16.

the surface of the triangle ABC equals that of the quadrilateral AFGC (Theorem 26).

If the point H coincides with the middle point E of the side BC (Fig. 17), only two equal right-angled triangles, ADF and BDE, are made, by whose interchange the equivalence of the surfaces of the triangle ABC and the quadrilateral AFEC is established.

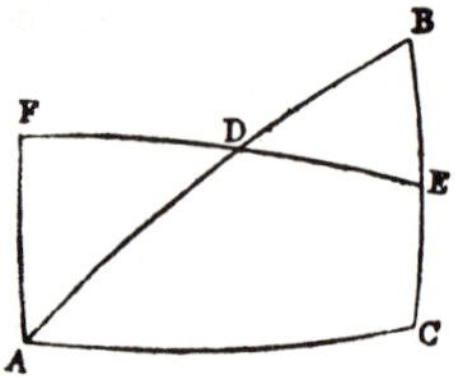

FIG. 17.

If, finally, the point H falls outside the triangle ABC (Fig. 18), the perpendicular CG goes, in consequence, through the triangle, and so we go over from the triangle ABC to the quadrilateral AFGC by adding the

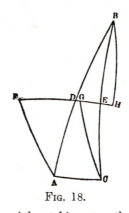

FIG. 18.

triangle FAD = DBH, and then taking away the triangle CGE = EBH.

Supposing in the spherical quadrilateral AFGC a great circle passed through the points A and G, as also through F and C, then will their arcs between AG and FC equal one another (Theorem 15), consequently also the triangles FAC and ACG be congruent (Theorem 15), and the angle FAC equal the angle ACG.

Hence follows, that in all the preceding cases, the sum of all three angles of the spherical triangle equals the sum of the two equal angles in the quadrilateral which are not the right angles.

Therefore we can, for every spherical triangle, in which the sum of the three angles is S, find a quadrilateral with equivalent surface, in which are two right angles and two equal perpendicular sides, and where the two other angles are each ½S.

Let now ABCD (Fig. 19) be the spherical quadrilateral, where the sides AB = DC are perpendicular to BC, and the angles A and D each $\frac{1}{2}$S.

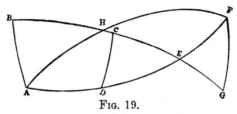

FIG. 19.

Prolong the sides AD and BC until they cut one another in E, and further beyond E, make DE = EF and let fall upon the prolongation of BC the perpendicular FG. Bisect the whole arc BG and join the mid-point H by great-circle-arcs with A and F.

The triangles EFG and DCE are congruent (Theorem 15), so FG = DC = AB.

The triangles ABH and HGF are likewise congruent, since they are right angled and have equal perpendicular sides, consequently AH and AF pertain to *one* circle, the arc AHF = π, ADEF likewise = π, the angle HAD = HFE = $\frac{1}{2}$S − BAH = $\frac{1}{2}$S − HFG = $\frac{1}{2}$S − HFE − EFG = $\frac{1}{2}$S − HAD − π + $\frac{1}{2}$S; consequently, angle HFE = $\frac{1}{2}$(S−π); or what is the same, this equals the size of the lune AHFDA, which again is equal to the quadrilateral ABCD, as we easily see if we pass over from the one to the other by first adding the triangle EFG and then BAH and thereupon taking away the triangles equal to them DCE and HFG.

Therefore $\frac{1}{2}$(S−π) is the size of the quadrilateral ABCD and at the same time also that of the spherical triangle in which the sum of the three angles is equal to S.

28. *If three planes cut each other in parallel lines, then the sum of the three surface angles equals two right angles.*

Let AA', BB' CC' (Fig. 20) be three parallels made by the inter-section of planes (Theorem 25). Take upon them at random three

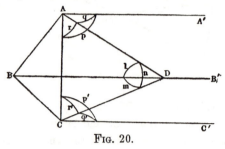

FIG. 20.

points A, B, C, and suppose through these a plane passed, which con-sequently will cut the planes of the parallels along the straight lines AB, AC, and BC. Further, pass through the line AC and any point D on the BB', another plane, whose intersection with the two planes of the parallels AA' and BB', CC' and BB' produces the two lines AD and DC, and whose inclination to the third plane of the parallels AA' and CC' we will designate by w.

The angles between the three planes in which the parallels lie will be designated by X, Y, Z, respectively at the lines AA', BB', CC'; finally call the linear angles BDC = a, ADC = b, ADB = c.

About A as center suppose a sphere described, upon which the inter-sections of the straight lines AC, AD AA' with it determine a spherical triangle, with the sides p, q, and r. Call its size α. Opposite the side q lies the angle w, opposite r lies X, and consequently opposite p lies the angle $\pi + 2\alpha - w - X$, (Theorem 27).

In like manner CA, CD, CC' cut a sphere about the center C, and determine a triangle of size β, with the sides p', q', r', and the angles, w opposite q', Z opposite r', and consequently $\pi + 2\beta - w - Z$ opposite p'.

Finally is determined by the intersection of a sphere about D with the lines DA, DB, DC, a spherical triangle, whose sides are l, m, n, and the angles opposite them $w + Z - 2\beta$, $w + X - 2\alpha$, and Y. Consequently its size $\delta = \frac{1}{2}(X + Y + Z - \pi) - \alpha - \beta + w$.

Decreasing w lessens also the size of the triangles α and β, so that $\alpha + \beta - w$ can be made smaller than any given number.

In the triangle δ can likewise the sides l and m be lessened even to vanishing (Theorem 21), consequently the triangle δ can be placed with one of its sides l or m upon a great circle of the sphere as often as you choose without thereby filling up the half of the sphere, hence δ vanishes together with w; whence follows that necessarily we must have

$$X+Y+Z = \pi$$

29. *In a rectilineal triangle, the perpendiculars erected at the mid-points of the sides either do not meet, or they all three cut each other in one point.*

Having pre-supposed in the triangle ABC (Fig. 21), that the two perpendiculars ED and DF, which are erected upon the sides AB and BC at their mid points E and F, intersect in the point D, then draw within the angles of the triangle the lines DA, DB, DC.

In the congruent triangles ADE and BDE (Theorem 10), we have AD = BD, thus follows also that BD = CD; the triangle ADC is hence isosceles, consequently the perpendicular dropped from the vertex D upon the base AC falls upon G the mid point of the base.

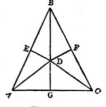

The proof remains unchanged also in the case when the intersection point D of the two perpendiculars ED and FD falls in the line AC itself, or falls without the triangle.

Fig. 21.

In case we therefore pre-suppose that two of those perpendiculars do not intersect, then also the third can not meet with them.

30. *The perpendiculars which are erected upon the sides of a rectilineal triangle at their mid-points, must all three be parallal to each other, so soon as the parallelism of two of them is pre-supposed.*

In the triangle ABC (Fig. 22) let the lines DE, FG, HK, be erected perpendicular upon the sides at their mid-points D, F, H. We will in the first place assume that the two perpendiculars DE and FG are parallel, cutting the line AB in L and M, and that the perpendicular HK lies between them. Within the angle BLE draw from the point L, at random, a straight line LG, which must cut FG somewhere in G,

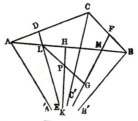

Fig. 22.

how small soever the angle of deviation GLE may be. (Theorem 16).

Since in the triangle LGM the perpendicular HK can not meet with MG (Theorem 29), therefore it must cut LG somewhere in P, whence follows, that HK is parallel to DE (Theorem 16), and to MG (Theorems 18 and 25).

Put the side $BC = 2a$, $AC = 2b$, $AB = 2c$, and designate the angles opposite these sides by A, B, C, then we have in the case just considered

$$A = \varPi(b) - \varPi(c),$$
$$B = \varPi(a) - \varPi(c),$$
$$C = \varPi(a) + \varPi(b),$$

as one may easily show with help of the lines AA', BB', CC', which are drawn from the points A, B, C, parallel to the perpendicular HK and consequently to both the other perpendiculars DE and FG (Theorems 23 and 25).

Let now the two perpendiculars HK and FG be parallel, then can the third DE not cut them (Theorem 29), hence is it either parallel to them, or it cuts AA'.

The last assumption is not other than that the angle

$$C > \varPi(a) + \varPi(b.)$$

If we lessen this angle, so that it becomes equal to $\varPi(a) + \varPi(b)$, while we in that way give the line AC the new position CQ, (Fig. 23), and designate the size of the third side BQ by $2c'$, then must the angle CBQ at the point B, which is increased, in accordance with what is proved above, be equal to

$$\varPi(a) - \varPi(c') > \varPi(a) - \varPi(c),$$

whence follows $c' > c$ (Theorem 23).

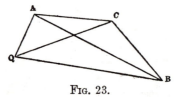

Fig. 23.

In the triangle ACQ are, however, the angles at A and Q equal, hence in the triangle ABQ must the angle at Q be greater than that at the point A, consequently is $AB > BQ$, (Theorem 9); that is $c > c'$.

31. *We call boundary line (oricycle) that curve lying in a plane for which all perpendiculars erected at the mid-points of chords are parallel to each other.*

In conformity with this definition we can represent the generation of a boundary line, if we draw to a given line AB (Fig. 24) from a given

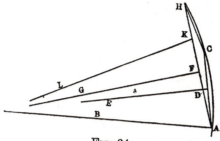

FIG. 24.

point A in it, making different angles CAB $= \Pi(a)$, chords AC $= 2a$; the end C of such a chord will lie on the boundary line, whose points we can thus gradually determine.

The perpendicular DE erected upon the chord AC at its mid-point D will be parallel to the line AB, which we will call the *Axis of the boundary line*. In like manner will also each perpendicular FG erected at the mid-point of any chord AH, be parallel to AB, consequently must this peculiarity also pertain to every perpendicular KL in general which is erected at the mid-point K of any chord CH, between whatever points C and H of the boundary line this may be drawn (Theorem 30). Such perpendiculars must therefore likewise, without distinction from AB, be called *Axes of the boundary line*.

32.   *A circle with continually increasing radius merges into the boundary line.*

Given AB (Fig. 25) a chord of the boundary line; draw from the end-points A and B of the chord two axes AC and BF, which consequently will make with the chord two equal angles BAC $=$ ABF $= \alpha$ (Theorem 31).

Upon one of these axes AC, take anywhere the point E as center of a circle, and draw the arc AF from the initial point A of the axis AC to its intersection point F with the other axis BF.

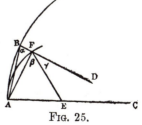

FIG. 25.

The radius of the circle, FE, corresponding to the point F will make on the one side with the chord AF an angle AFE $= \beta$, and on the other side with the axis BF, the angle $EFD = \gamma$. It follows that the angle between the two chords $BAF = a - \beta < \beta + \gamma - a$ (Theorem 22); whence follows, $a - \beta < \frac{1}{2}\gamma$.

Since now however the angle γ approaches the limit zero, as well in consequence of a moving of the center E in the direction AC, when F remains unchanged, (Theorem 21), as also in consequence of an approach of F to B on the axis BF, when the center E remains in its position (Theorem 22), so it follows, that with such a lessening of the angle γ, also the angle $a - \beta$, or the mutual inclination of the two chords AB and AF, and hence also the distance of the point B on the boundary line from the point F on the circle, tends to vanish.

Consequently one may also call the boundary-line *a circle with infinitely great radius.*

33.   Let $AA' = BB' = x$ (Figure 26), be two lines parallel toward the side from A to A', which parallels serve as axes for the two boundary arcs (arcs on two boundary lines) $AB = s$, $A'B' = s'$, then is

$$s' = s e^{-x}$$

where e is independent of the arcs s, s' and of the straight line x, the distance of the arc s' from s.

Fig. 26.

In order to prove this, assume that the ratio of the arc s to s' is equal to the ratio of the two whole numbers n and m.

Between the two axes AA', BB' draw yet a third axis CC', which so cuts off from the arc AB a part $AC = t$ and from the arc A'B' on the same side, a part $A'C' = t'$. Assume the ratio of t to s equal to that of the whole numbers p and q, so that

$$s = \frac{n}{m} s', \quad t = \frac{p}{q} s.$$

Divide now s by axes into nq equal parts, then will there be mq such parts on s' and np on t.

However there correspond to these equal parts on s and t likewise equal parts on s' and t', consequently we have

$$\frac{t'}{t} = \frac{s'}{s}$$

Hence also wherever the two arcs t and t' may be taken between the two axes AA' and BB', the ratio of t to t' remains always the same, as

long as the distance x between them remains the same. If we therefore for $x = 1$, put $s = es'$, then we must have for every x

$$s' = se^{-x}.$$

Since e is an unknown number only subjected to the condition $e > 1$,
and further the linear unit for x may be taken at will, therefore we may,
for the simplification of reckoning, so choose it that by e is to be understood the base of Napierian logarithms.

We may here remark, that $s' = 0$ for $x = \infty$, hence not only does
the distance between two parallels decrease (Theorem 24), but with the
prolongation of the parallels toward the side of the parallelism this at
last wholly vanishes. Parallel lines have therefore the character of
asymptotes.

34. *Boundary surface* (*orisphere*) we call that surface which arises
from the revolution of the boundary line about one of its axes, which,
together with all other axes of the boundary-line, will be also an axis
of the boundary-surface.

*A chord is inclined at equal angles to such axes drawn through its end-
points, wheresoever these two end-points may be taken on the boundary-surface.*

Let A, B, C, (Fig. 27), be three points on the boundary-surface;

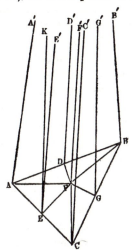

Fig. 27.

AA', the axis of revolution, BB' and CC' two other axes, hence **AB**
and AC chords to which the axes are inclined at equal angles A'AB
$=$ B'BA, A'AC $=$ C'CA (Theorem 31.)

Two axes BB', CC', drawn through the end-points of the third chord BC, are likewise parallel and lie in one plane, (Theorem 25).

A perpendicular DD' erected at the mid-point D of the chord AB and in the plane of the two parallels AA', BB', must be parallel to the three axes AA', BB', CC', (Theorems 23 and 25); just such a perpendicular EE' upon the chord AC in the plane of the parallels AA', CC' will be parallel to the three axes AA', BB', CC', and the perpendicular DD'. Let now the angle between the plane in which the parallels AA' and BB' lie, and the plane of the triangle ABC be designated by $\Pi(a)$, where a may be positive, negative or null. If a is positive, then erect FD $= a$ within the triangle ABC, and in its plane, perpendicular upon the chord AB at its mid-point D.

Were a a negative number, then must FD $= a$ be drawn outside the triangle on the other side of the chord AB; when $a = 0$, the point F coincides with D.

In all cases arise two congruent right-angled triangles AFD and DFB, consequently we have FA $=$ FB.

Erect now at F the line FF' perpendicular to the plane of the triangle ABC.

Since the angle D'DF $= \Pi(a)$, and DF $= a$, so FF' is parallel to DD' and the line EE', with which also it lies in one plane perpendicular to the plane of the triangle ABC.

Suppose now in the plane of the parallels EE', FF' upon EF the perpendicular EK erected, then will this be also at right angles to the plane of the triangle ABC, (Theorem 13), and to the line AE lying in this plane, (Theorem 11); and consequently must AE, which is perpendicular to EK and EE', be also at the same time perpendicular to FE, (Theorem 11). The triangles AEF and FEC are congruent, since they are right-angled and have the sides about the right angles equal, hence is
$$AF = FC = FB.$$
A perpendicular from the vertex F of the isosceles triangle BFC let fall upon the base BC, goes through its mid-point G; a plane passed through this perpendicular FG and the line FF' must be perpendicular to the plane of the triangle ABC, and cuts the plane of the parallels BB', CC', along the line GG', which is likewise parallel to BB' and CC', (Theorem 25); since now CG is at right angles to FG, and hence at the same time also to GG', so consequently is the angle C'CG $=$ B'BG, (Theorem 23).

Hence follows, that for the boundary-surface each of the axes may be considered as axis of revolution.

Principal-plane we will call each plane passed through an axis of the boundary surface.

Accordingly every *Principal-plane* cuts the boundary-surface in the boundary line, while for another position of the cutting plane this intersection is a circle.

Three principal planes which mutually cut each other, make with each other angles whose sum is π, (Theorem 28).

These angles we will consider as angles in the boundary-triangle whose sides are arcs of the boundary-line, which are made on the boundary surface by the intersections with the three principal planes. Consequently the same interdependence of the angles and sides pertains to the boundary-triangles, that is proved in the ordinary geometry for the rectilineal triangle.

35. In what follows, we will designate the size of a line by a letter with an accent added, *e. g.* x', in order to indicate that this has a relation to that of another line, which is represented by the same letter without accent x, which relation is given by the equation

$$\Pi(x) + \Pi(x') = \tfrac{1}{2}\pi.$$

Let now ABC (Fig. 28) be a rectilineal right-angled triangle, where the hypothenuse AB $= c$, the other sides AC $= b$, BC $= a$, and the

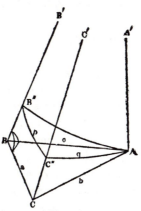

Fig. 28.

angles opposite them are

$$BAC = \Pi(\alpha), \quad ABC = \Pi(\beta).$$

At the point A erect the line AA' at right angles to the plane of the triangle ABC, and from the points B and C draw BB' and CC' parallel to AA'.

The planes in which these three parallels lie make with each other the angles: $\Pi(a)$ at AA', a right angle at CC' (Theorems 11 and 13), consequently $\Pi(a')$ at BB' (Theorem 28).

The intersections of the lines BA, BC, BB' with a sphere described about the point B as center, determine a spherical triangle mnk, in which the sides are $mn = \Pi(c)$, $kn = \Pi(\beta)$, $mk = \Pi(a)$ and the opposite angles are $\Pi(b)$, $\Pi(a')$, $\frac{1}{2}\pi$.

Therefore we must, with the existence of a rectilineal triangle whose sides are a, b, c and the opposite angles $\Pi(a)$, $\Pi(\beta)$ $\frac{1}{2}\pi$, also admit the existence of a spherical triangle (Fig. 29) with the sides $\Pi(c)$, $\Pi(\beta)$, $\Pi(a)$ and the opposite angles $\Pi(b)$, $\Pi(a')$, $\frac{1}{2}\pi$.

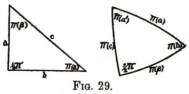

Fig. 29.

Of these two triangles, however, also inversely the existence of the spherical triangle necessitates anew that of a rectilineal, which in consequence, also can have the sides a, a', β, and the opposite angles $\Pi(b')$, $\Pi(c)$, $\frac{1}{2}\pi$.

Hence we may pass over from a, b, c, a, β, to b, a, c, β, a, and also to a, a', β, b', c.

Suppose through the point A (Fig. 28) with AA' as axis, a boundary-surface passed, which cuts the two other axes BB', CC', in B'' and C'', and whose intersections with the planes the parallels form a boundary-triangle, whose sides are B''C'' = p, C''A = q, B''A = r, and the angles opposite them $\Pi(a)$, $\Pi(a')$, $\frac{1}{2}\pi$, and where consequently (Theorem 34):

$$p = r \sin \Pi(a), \quad q = r \cos \Pi(a).$$

Now break the connection of the three principal-planes along the line BB', and turn them out from each other so that they with all the lines lying in them come to lie in one plane, where consequently the arcs p, q, r will unite to a single arc of a boundary-line, which goes through the point A and has AA′ for axis, in such a manner that (Fig. 30) on the one side will lie, the arcs q and p, the side b of the triangle, which is

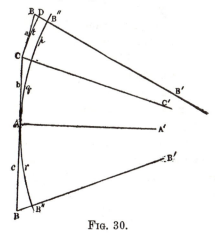

FIG. 30.

perpendicular to AA′ at A, the axis CC′ going from the end of b parallel to AA′ and through C″ the union point of p and q, the side a perpendicular to CC′ at the point C, and from the end-point of a the axis BB′ parallel to AA′ which goes through the end-point B″ of the arc p.

On the other side of AA′ will lie, the side c perpendicular to AA′ at the point A, and the axis BB′ parallel to AA′, and going through the end-point B″ of the arc r remote from the end point of b.

The size of the line CC″ depends upon b, which dependence we will express by $CC″ = f(b)$.

In like manner we will have $BB″ = f(c)$.

If we describe, taking CC′ as axis, a new boundary line from the point C to its intersection D with the axis BB′ and designate the arc CD by t, then is $BD = f(a)$.

$$BB′ = BD + DB″ = BD + CC″, \text{ consequently}$$
$$f(c) = f(a) + f(b).$$

Moreover, we perceive, that (Theorem 33)

$$t = p e^{f(b)} = r \sin \varPi(a) \, e^{f(b)}.$$

If the perpendicular to the plane of the triangle ABC (Fig. 28) were erected at B instead of at the point A, then would the lines c and r remain the same, the arcs q and t would change to t and q, the straight lines a and b into b and a, and the angle $\Pi(\alpha)$ into $\Pi(\beta)$, consequently we would have

$$q = r \sin \Pi(\beta) \, e^{f(a)},$$

whence follows by substituting the value of q,

$$\cos \Pi(\alpha) = \sin \Pi(\beta) \, e^{f(a)},$$

and if we change α and β into b' and c,

$$\sin \Pi(b) = \sin \Pi(c) \, e^{f(a)};$$

further, by multiplication with $e^{f(b)}$

$$\sin \Pi(b) \, e^{f(b)} = \sin \Pi(c) \, e^{f(c)}$$

Hence follows also

$$\sin \Pi(a) \, e^{f(a)} = \sin \Pi(b) \, e^{f(b)}.$$

Since now, however, the straight lines a and b are independent of one another, and moreover, for b=0, $f(b)=0$, $\Pi(b)=\tfrac{1}{2}\pi$, so we have for every straight line a

$$e^{-f(a)} = \sin \Pi(a).$$

Therefore,

$$\sin \Pi(c) = \sin \Pi(a) \sin \Pi(b),$$
$$\sin \Pi(\beta) = \cos \Pi(\alpha) \sin \Pi(a).$$

Hence we obtain besides by mutation of the letters

$$\sin \Pi(\alpha) = \cos \Pi(\beta) \sin \Pi(b),$$
$$\cos \Pi(b) = \cos \Pi(c) \cos \Pi(\alpha),$$
$$\cos \Pi(a) = \cos \Pi(c) \cos \Pi(\beta).$$

If we designate in the right-angled spherical triangle (Fig. 29) the sides $\Pi(c)$, $\Pi(\beta)$, $\Pi(a)$, with the opposite angles $\Pi(b)$, $\Pi(a')$, by the letters a, b, c, A, B, then the obtained equations take on the form of those which we know as proved in spherical trigonometry for the right-angled triangle, namely,

$$\sin a = \sin c \sin A,$$
$$\sin b = \sin c \sin B,$$
$$\cos A = \cos a \sin B,$$
$$\cos B = \cos b, \sin A,$$
$$\cos c = \cos a, \cos b;$$

from which equations we can pass over to those for all spherical triangles in general.

Hence spherical trigonometry is not dependent upon whether in a rectilineal triangle the sum of the three angles is equal to two right angles or not.

36.   We will now consider anew the right-angled rectilineal triangle ABC (Fig. 31), in which the sides are a, b, c, and the opposite angles $\Pi(a)$, $\Pi(\beta)$, $\tfrac{1}{2}\pi$.

Prolong the hypothenuse c through the point B, and make BD$=\beta$; at the point D erect upon BD the perpendicular DD', which consequently will be parallel to BB', the prolongation of the side a beyond the point B. Parallel to DD' from the point A draw AA', which is at the same time also parallel to CB', (Theorem 25), therefore is the angle

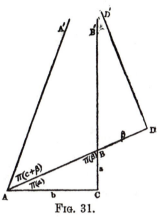

$$A'AD = \Pi(c+\beta),$$
$$A'AC = \Pi(b), \text{ consequently}$$
$$\Pi(b) = \Pi(a) + \Pi(c+\beta).$$

FIG. 31.

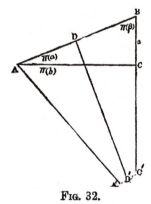

FIG. 32.

If from B we lay off β on the hypothenuse c, then at the end point D, (Fig. 32), within the triangle erect upon AB the perpendicular DD', and from the point A parallel to DD' draw AA', so will BC with its prolongation CC' be the third parallel; then is, angle CAA' $=\Pi$ (b), DAA' $=\Pi(c-\beta)$, consequently $\Pi(c-\beta) = \Pi(a) + \Pi(b)$. The last equation is then also still valid, when $c=\beta$, or $c<\beta$.

If $c=\beta$ (Fig. 33), then the perpendicular AA' erected upon AB at the point A is parallel to the side BC=a, with its prolongation, CC′, consequently

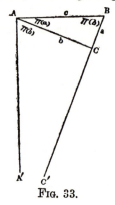

Fig. 33.

we have $\Pi(a)+\Pi(b)=\frac{1}{2}\pi$, whilst also $\Pi(c-\beta)=\frac{1}{2}\pi$, (Theorem 23).

If $c<\beta$, then the end of β falls beyond the point A at D (Fig. 34) upon the prolongation of the hypothenuse AB. Here the perpendicular DD′ erected upon AD, and the line AA′ parallel to it from A, will

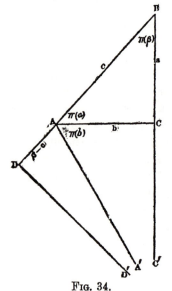

Fig. 34.

likewise be parallel to the side BC=a, with its prolongation CC′.

Here we have the angle DAA′ $= \Pi$ $(\beta-c)$, consequently
$\Pi(a)+\Pi(b)=\pi-\Pi(\beta-c)=\Pi(c-\beta)$, (Theorem 23).

The combination of the two equations found gives,

$$2\,\Pi(b)=\Pi(c-\beta)+\Pi(c+\beta),$$
$$2\,\Pi(a)=\Pi(c-\beta)-\Pi(c+\beta),$$

whence follows

$$\frac{\cos \Pi(b)}{\cos \Pi(a)}=\frac{\cos\left[\frac{1}{2}\Pi(c-\beta)+\frac{1}{2}\,\Pi(c+\beta)\right]}{\cos\left[\frac{1}{2}\Pi(c-\beta)-\frac{1}{2}\,\Pi(c+\beta)\right]}$$

Substituting here the value, (Theorem 35)

$$\frac{\cos \Pi(b)}{\cos \Pi(a)}=\cos\Pi(c),$$

we have $[\tan\frac{1}{2}\Pi(c)]^2=\tan\frac{1}{2}\Pi(c-\beta)\tan\frac{1}{2}\Pi(c+\beta).$

Since here β is an arbitrary number, as the angle $\Pi(\beta)$ at the one side of c may be chosen at will between the limits 0 and $\frac{1}{2}\pi$, consequently β between the limits 0 and ∞, so we may deduce by taking consecutively $\beta = c$, 2c, 3c, &c., that for every positive number n, $[\tan\frac{1}{2}\Pi(c)]^n = \tan\frac{1}{2}\Pi(nc)$.

If we consider n as the ratio of two lines x and c, and assume that $\cot\frac{1}{2}\Pi(c) = e^c$,

then we find for every line x in general, whether it be positive or negative, $\tan\frac{1}{2}\Pi(x) = e^{-x}$

where e may be any arbitrary number, which is greater than unity, since $\Pi(x) = 0$ for $x = \infty$.

Since the unit by which the lines are measured is arbitrary, so we may also understand by e the base of the Napierian Logarithms.

37. Of the equations found above in Theorem 35 it is sufficient to know the two following,

$$\sin \Pi(c) = \sin \Pi(a) \sin \Pi(b)$$
$$\sin \Pi(a) = \sin \Pi(b) \cos \Pi(\beta),$$

applying the latter to both the sides a and b about the right angle, in order from the combination to deduce the remaining two of Theorem 35, without ambiguity of the algebraic sign, since here all angles are acute.

In a similar manner we attain the two equations

(1.) $\quad \tan \Pi(c) = \sin \Pi(a) \tan \Pi(a)$,

(2.) $\quad \cos \Pi(a) = \cos \Pi(c) \cos \Pi(\beta)$.

We will now consider a rectilineal triangle whose sides are a, b, c, (Fig. 35) and the opposite angles A, B, C.

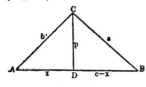

If A and B are acute angles, then the perpendicular p from the vertex of the angle C falls within the triangle and cuts the side c into two parts, x on the side of the angle A and c—x on the side of the angle B. Thus arise two right-angled triangles, for which we obtain, by application of equation (1),

Fig. 35.

$$\tan \Pi(a) = \sin B \tan \Pi(p),$$
$$\tan \Pi(b) = \sin A \tan \Pi(p),$$

which equations remain unchanged also when one of the angles, *e. g.* B, is a right angle (Fig. 36) or and obtuse angle (Fig. 37).

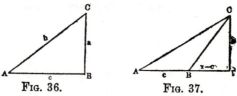

Fig. 36.   Fig. 37.

Therefore we have universally for every triangle

(3.)   sin A tan Π(a)=sin B tan Π(b).

For a triangle with acute angles A, B, (Fig. 35) we have also (Equation 2),

$$\cos \Pi(x)=\cos A \cos \Pi(b),$$
$$\cos \Pi(c-x)=\cos B \cos \Pi(a)$$

which equations also relate to triangles, in which one of the angles A or B is a right angle or an obtuse angle.

As example, for B=$\frac{1}{2}\pi$ (Fig. 36) we must take x=c, the first equation then goes over into that which we have found above as Equation 2, the other, however, is self-sufficing.

For B>$\frac{1}{2}\pi$ (Fig. 37) the first equation remains unchanged, instead of the second, however, we must write correspondingly

$$\cos \Pi(x-c)=\cos (\pi-B) \cos \Pi(a);$$

but we have $\cos \Pi(x-c)=-\cos \Pi(c-x)$

(Theorem 23), and also cos $(\pi-B)=-\cos$ B.

If A is a right or an obtuse angle, then must c$-$x and x be put for x and c$-$x, in order to carry back this case upon the preceding.

In order to eliminate x from both equations, we notice that (Theorem 36)

$$\cos \Pi(c-x)= \frac{1-[\tan\frac{1}{2} \Pi(c-x)]^2}{1+[\tan \frac{1}{2}\Pi(c-x)]^2}$$
$$=\frac{1-e^{2x-2c}}{1+e^{2x-2c}}=$$
$$=\frac{1-[\tan \frac{1}{2}\Pi(c)]^2[\cot \frac{1}{2}\Pi(x)]^2}{1+[\tan \frac{1}{2}\Pi(c)]^2[\cot \frac{1}{2}\Pi(x)]^2}$$
$$=\frac{\cos \Pi(c)-\cos\Pi(x)}{1-\cos \Pi(c)\cos\Pi(x)}$$

If we substitute here the expression for $\cos \varPi(x)$, $\cos \varPi(c-x)$, we obtain

$$\cos \varPi(c) = \frac{\cos \varPi(a) \cos B + \cos \varPi(b) \cos A}{1 + \cos \varPi(a) \cos \varPi(b) \cos A \cos B}$$

whence follows

$$\cos \varPi(a) \cos B = \frac{\cos \varPi(c) - \cos A \cos \varPi(b)}{1 - \cos A \cos \varPi(b) \cos \varPi(c)}$$

and finally

$$[\sin \varPi(c)]^2 = [1 - \cos B \cos \varPi(c) \cos \varPi(a)][1 - \cos A \cos \varPi(b) \cos \varPi(c)]$$

In the same way we must also have

(4.)

$$[\sin \varPi(a)]^2 = [1 - \cos C \cos \varPi(a) \cos \varPi(b)][1 - \cos B \cos \varPi(c) \cos \varPi(a)]$$
$$[\sin \varPi(b)]^2 = [1 - \cos A \cos \varPi(b) \cos \varPi(c)][1 - \cos C \cos \varPi(a) \cos \varPi(b)]$$

From these three equations we find

$$\frac{[\sin \varPi(b)]^2 [\sin \varPi(c)]^2}{[\sin \varPi(a)]^2} = [1 - \cos A \cos \varPi(b) \cos \varPi(c)]^2.$$

Hence follows without ambiguity of sign,

(5.) $\qquad \cos A \cos \varPi(b) \cos \varPi(c) + \dfrac{\sin \varPi(b) \sin \varPi(c)}{\sin \varPi(a)} = 1.$

If we substitute here the value of $\sin \varPi(c)$ corresponding to equation (3.)

$$\sin \varPi(c) = \frac{\sin A}{\sin C} \tan \varPi(a) \cos \varPi(c)$$

then we obtain

$$\cos \varPi(c) = \frac{\cos \varPi(a) \sin C}{\sin A \sin \varPi(b) + \cos A \sin C \cos \varPi(a) \cos \varPi(b)};$$

but by substituting this expression for $\cos \varPi(c)$ in equation (4),

(6.) $\qquad \cot A \sin C \sin \varPi(b) + \cos C = \dfrac{\cos \varPi(b)}{\cos \varPi(a)}$

By elimination of $\sin \varPi(b)$ with help of the equation (3) comes

$$\frac{\cos \varPi(a)}{\cos \varPi(b)} \cos C = 1 - \frac{\cos A}{\sin B} \sin C \sin \varPi(a).$$

In the meantime the equation (6) gives by changing the letters,

$$\frac{\cos \varPi(a)}{\cos \varPi(b)} = \cot B \sin C \sin \varPi(a) + \cos C.$$

From the last two equations follows,

(7.) $$\cos A + \cos B \cos C = \frac{\sin B \sin C}{\sin \varPi(a)}$$

All four equations for the interdependence of the sides a, b, c, and the opposite angles A, B, C, in the rectilineal triangle will therefore be, [Equations (3), (5), (6), (7).]

(8.)
$$\begin{cases} \sin A \tan \varPi(a) = \sin B \tan \varPi(b), \\[2mm] \cos A \cos \varPi(b) \cos \varPi(c) + \dfrac{\sin \varPi(b) \sin \varPi(c)}{\sin \varPi(a)} = 1, \\[2mm] \cot A \sin C \sin \varPi(b) + \cos C = \dfrac{\cos \varPi(b)}{\cos \varPi(a)}, \\[2mm] \cos A + \cos B \cos C = \dfrac{\sin B \sin C}{\sin \varPi(a)}. \end{cases}$$

If the sides a, b, c, of the triangle are very small, we may content ourselves with the approximate determinations. (Theorem 36.)

$$\cot \varPi(a) = a,$$
$$\sin \varPi(a) = 1 - \tfrac{1}{2}a^2$$
$$\cos \varPi(a) = a,$$

and in like manner also for the other sides b and c.

The equations 8 pass over for such triangles into the following:

$$b \sin A = a \sin B,$$
$$a^2 = b^2 + c^2 - 2bc \cos A,$$
$$a \sin (A + C) = b \sin A,$$
$$\cos A + \cos (B + C) = 0.$$

Of these equations the first two are assumed in the ordinary geometry; the last two lead, with the help of the first, to the conclusion

$$A + B + C = \pi.$$

Therefore the imaginary geometry passes over into the ordinary, when we suppose that the sides of a rectilineal triangle are very small.

I have, in the scientific bulletins of the University of Kasan, published certain researches in regard to the measurement of curved lines, of plane figures, of the surfaces and the volumes of solids, as well as in relation to the application of imaginary geometry to analysis.

The equations (8) attain for themselves already a sufficient foundation for considering the assumption of imaginary geometry as possible. Hence there is no means, other than astronomical observations, to use for judging of the exactitude which pertains to the calculations of the ordinary geometry.

This exactitude is very far-reaching, as I have shown in one of my investigations, so that, for example, in triangles whose sides are attainable for our measurement, the sum of the three angles is not indeed different from two right-angles by the hundredth part of a second.

In addition, it is worthy of notice that the four equations (8) of plane geometry pass over into the equations for spherical triangles, if we put a $\sqrt{-1}$, b $\sqrt{-1}$, c $\sqrt{-1}$, instead of the sides a, b, c; with this change, however, we must also put

$$\sin \varPi (a) = \frac{1}{\cos (a),}$$
$$\cos \varPi (a) = (\sqrt{-1}) \tan a,$$
$$\tan \varPi (a) = \frac{1}{\sin a (\sqrt{-1}),}$$

and similarly also for the sides b and c.

In this manner we pass over from equations (8) to the following:

$$\sin A \sin b = \sin B \sin a,$$
$$\cos a = \cos b \cos c + \sin b \sin c \cos A,$$
$$\cot A \sin C + \cos C \cos b = \sin b \cot a,$$
$$\cos A = \cos a \sin B \sin C - \cos B \cos C.$$

TRANSLATOR'S APPENDIX.

ELLIPTIC GEOMETRY.

Gauss himself never published aught upon this fascinating subject, Geometry Non-Euclidean; but when the most extraordinary pupil of his long teaching life came to read his inaugural dissertation before the Philosophical Faculty of the University of Goettingen, from the three themes submitted it was the choice of Gauss which fixed upon the one "Ueber die Hypothesen welche der Geometrie zu Grunde liegen."

Gauss was then recognized as the most powerful mathematician in the world. I wonder if he saw that here his pupil was already beyond him, when in his sixth sentence Riemann says, "therefore space is only a special case of a three-fold extensive magnitude," and continues: "From this, however, it follows of necessity, that the propositions of geometry can not be deduced from general magnitude-ideas, but that those peculiarities through which space distinguishes itself from other thinkable threefold extended magnitudes can only be gotten from experience. Hence arises the problem, to find the simplest facts from which the metrical relations of space are determinable — a problem which from the nature of the thing is not fully determinate; for there may be obtained several systems of simple facts which suffice to determine the metrics of space; that of Euclid as weightiest is for the present aim made fundamental. These facts are, as all facts, not necessary, but only of empirical certainty; they are hypotheses. Therefore one can investigate their probability, which, within the limits of observation, of course is very great, and after this judge of the allowability of their extension beyond the bounds of observation, as well on the side of the immeasurably great as on the side of the immeasurably small."

Riemann extends the idea of curvature to spaces of three and more dimensions. The curvature of the sphere is constant and positive, and on it figures can freely move without deformation. The curvature of the plane is constant and zero, and on it figures slide without stretching. The curvature of the two-dimentional space of Lobachevski and

Bolyai completes the group, being constant and negative, and in it figures can move without stretching or squeezing. As thus corresponding to the sphere it is called the pseudo-sphere.

In the space in which we live, we suppose we can move without deformation. It would then, according to Riemann, be a special case of a space of constant curvature. We presume its curvature null. At once the supposed fact that our space does not interfere to squeeze us or stretch us when we move, is envisaged as a peculiar property of our space. But is it not absurd to speak of space as interfering with anything? If you think so, take a knife and a raw potato, and try to cut it into a seven-edged solid.

Further on in this astonishing discourse comes the epoch-making idea, that though space be unbounded, it is not therefore infinitely great. Riemann says: "In the extension of space-constructions to the immeasurably great, the unbounded is to be distinguished from the infinite; the first pertains to the relations of extension, the latter to the size-relations.

"That our space is an unbounded three-fold extensive manifoldness, is a hypothesis, which is applied in each apprehension of the outer world, according to which, in each moment, the domain of actual perception is filled out, and the possible places of a sought object constructed, and which in these applications is continually confirmed. The unboundedness of space possesses therefore a greater empirical certainty than any outer experience. From this however the Infinity in no way follows. Rather would space, if one presumes bodies independent of place, that is ascribes to it a constant curvature, necessarily be finite so soon as this curvature had ever so small a positive value. One would, by extending the beginnings of the geodesics lying in a surface-element, obtain an unbounded surface with constant positive curvature, therefore a surface which in a homaloidal three-fold extensive manifoldness would take the form of a sphere, and so is finite."

Here we have for the first time in human thought the marvelous perception that universal space may yet be only finite.

Assume that a straight line is uniquely determined by two points, but take the contradictory of the axiom that a straight line is of infinite size; then the straight line returns into itself, and two having intersected get back to that intersection point.

BIBLIOGRAPHY.

A bibliography of non-Euclidean literature down to the year 1878 was given by Halsted, "*American Journal of Mathematics*," vols. i, ii, containing 81 authors and 174 titles, and reprinted in the collected works of Lobachevski (Kazan, 1886) giving 124 authors and 272 titles. This was incorporated in Bonola's Bibliography of the Foundations of Geometry (1899) reprinted (1902) at Kolozsvár in the Bolyai Memorial Volume. In 1911 appeared the volume: Bibliography of Non-Euclidean Geometry by Duncan M. Y. Sommerville; London, Harrison and Sons.

The Introduction says: "The present work was begun about nine years ago. It was intended as a continuation of Halsted's bibliography, but it soon became evident that the growth of the subject rendered such diffuse treatment practically impossible, and short abstracts of the works would have to be dispensed with. The object is to produce as far as possible a complete repository of the titles of all works from the earliest times up to the present which deal with the extended conception of space, and to form a guide to the literature in an easily accessible form. It includes the theory of parallels, non-euclidean geometry, the foundations of geometry, and space of n dimensions."

In 1913 Teubner issued in two parts Paul Stäckel's important book: Wolfgang und Johann Bolyai. Geometrische Untersuchungen. John compares Lobachevski's researches with his own. The profound philosophic import of non-euclidean geometry forms an integrant part of "The Foundations of Science," by H. Poincaré; Vol. I of the series Science and Education, The Science Press, New York City, 1914. The Transactions of the Royal Society of Canada, Vol. XII, Section III, contains a striking Presidential Address by Alfred Baker on The Foundations of Geometry. Of the cognate works issued by The Open Court Pub. Co., we mention only Euclid's Parallel Postulate by Withers. Scores of errors are pointed out in "Non-Euclidean Geometry in the Encyclopædia Britannica," Science, May 10, 1912.

And now at last the theory of relativity has made non-euclidean geometry a powerful machine for advance in physics.

Says Vladimir Varićak in a remarkable lecture, "Ueber die nichteuklidische Interpretation der Relativtheorie," (Jahresber. D. Math. Ver., 21, 103-127),

I postulated that the phenomena happened in a Lobachevski space, and reached by very simple geometric deduction the formulas of the relativity theory. Assuming non-euclidean terminology, the formulas of the relativity theory become not only essentially simplified, but capable of a geometric interpretation wholly analogous to the interpretation of the classic theory in the euclidean geometry. And this analogy often goes so far, that the very wording of the theorems of the classic theory may be left unchanged.

A CATALOG OF SELECTED
DOVER BOOKS
IN SCIENCE AND MATHEMATICS

A CATALOG OF SELECTED
DOVER BOOKS
IN SCIENCE AND MATHEMATICS

QUALITATIVE THEORY OF DIFFERENTIAL EQUATIONS, V.V. Nemytskii and V.V. Stepanov. Classic graduate-level text by two prominent Soviet mathematicians covers classical differential equations as well as topological dynamics and erqodic theory. Bibliographies. 523pp. 5⅜ × 8½. 65954-2 Pa. $10.95

MATRICES AND LINEAR ALGEBRA, Hans Schneider and George Phillip Barker. Basic textbook covers theory of matrices and its applications to systems of linear equations and related topics such as determinants, eigenvalues and differential equations. Numerous exercises. 432pp. 5⅜ × 8½. 66014-1 Pa. $8.95

QUANTUM THEORY, David Bohm. This advanced undergraduate-level text presents the quantum theory in terms of qualitative and imaginative concepts, followed by specific applications worked out in mathematical detail. Preface. Index. 655pp. 5⅜ × 8½. 65969-0 Pa. $10.95

ATOMIC PHYSICS (8th edition), Max Born. Nobel laureate's lucid treatment of kinetic theory of gases, elementary particles, nuclear atom, wave-corpuscles, atomic structure and spectral lines, much more. Over 40 appendices, bibliography. 495pp. 5⅜ × 8½. 65984-4 Pa. $11.95

ELECTRONIC STRUCTURE AND THE PROPERTIES OF SOLIDS: The Physics of the Chemical Bond, Walter A. Harrison. Innovative text offers basic understanding of the electronic structure of covalent and ionic solids, simple metals, transition metals and their compounds. Problems. 1980 edition. 582pp. 6⅛ × 9¼. 66021-4 Pa. $14.95

BOUNDARY VALUE PROBLEMS OF HEAT CONDUCTION, M. Necati Özisik. Systematic, comprehensive treatment of modern mathematical methods of solving problems in heat conduction and diffusion. Numerous examples and problems. Selected references. Appendices. 505pp. 5⅜ × 8½. 65990-9 Pa. $11.95

A SHORT HISTORY OF CHEMISTRY (3rd edition), J.R. Partington. Classic exposition explores origins of chemistry, alchemy, early medical chemistry, nature of atmosphere, theory of valency, laws and structure of atomic theory, much more. 428pp. 5⅜ × 8½. (Available in U.S. only) 65977-1 Pa. $10.95

A HISTORY OF ASTRONOMY, A. Pannekoek. Well-balanced, carefully reasoned study covers such topics as Ptolemaic theory, work of Copernicus, Kepler, Newton, Eddington's work on stars, much more. Illustrated. References. 521pp. 5⅜ × 8½. 65994-1 Pa. $11.95

PRINCIPLES OF METEOROLOGICAL ANALYSIS, Walter J. Saucier. Highly respected, abundantly illustrated classic reviews atmospheric variables, hydrostatics, static stability, various analyses (scalar, cross-section, isobaric, isentropic, more). For intermediate meteorology students. 454pp. 6⅛ × 9¼. 65979-8 Pa. $12.95

SPECIAL FUNCTIONS, N.N. Lebedev. Translated by Richard Silverman. Famous Russian work treating more important special functions, with applications to specific problems of physics and engineering. 38 figures. 308pp. 5⅜ × 8½.
60624-4 Pa. $6.95

OBSERVATIONAL ASTRONOMY FOR AMATEURS, J.B. Sidgwick. Mine of useful data for observation of sun, moon, planets, asteroids, auroiae, meteors, comets, variables, binaries, etc. 39 illustrations 384pp. 5⅜ × 8¼. (Available in U.S. only)
24033-9 Pa. $5.95

INTEGRAL EQUATIONS, F.G. Tricomi. Authoritative, well-written treatment of extremely useful mathematical tool with wide applications. Volterra Equations, Fredholm Equations, much more. Advanced undergraduate to graduate level. Exercises. Bibliography. 238pp. 5⅜ × 8½.
64828-1 Pa. $6.95

CELESTIAL OBJECTS FOR COMMON TELESCOPES, T.W. Webb. Inestimable aid for locating and identifying nearly 4,000 celestial objects. 77 illustrations. 645pp. 5⅜ × 8½.
20917-2, 20918-0 Pa., Two-vol. set $12.00

MODERN NONLINEAR EQUATIONS, Thomas L. Saaty. Emphasizes practical solution of problems; covers seven types of equations. ". . . a welcome contribution to the existing literature. . . ."—*Math Reviews.* 490pp. 5⅜ × 8½. 64232-1 Pa. $9.95

FUNDAMENTALS OF ASTRODYNAMICS, Roger Bate et al. Modern approach developed by U.S. Air Force Academy. Designed as a first course. Problems, exercises. Numerous illustrations. 455pp. 5⅜ × 8½.
60061-0 Pa. $8.95

INTRODUCTION TO LINEAR ALGEBRA AND DIFFERENTIAL EQUATIONS, John W. Dettman. Excellent text covers complex numbers, determinants, orthonormal bases, Laplace transforms, much more. Exercises with solutions. Undergraduate level. 416pp. 5⅜ × 8½.
65191-6 Pa. $8.95

INCOMPRESSIBLE AERODYNAMICS, edited by Bryan Thwaites. Covers theoretical and experimental treatment of the uniform flow of air and viscous fluids past two-dimensional aerofoils and three-dimensional wings; many other topics. 654pp. 5⅜ × 8½.
65465-6 Pa. $14.95

INTRODUCTION TO DIFFERENCE EQUATIONS, Samuel Goldberg. Exceptionally clear exposition of important discipline with applications to sociology, psychology, economics. Many illustrative examples; over 250 problems. 260pp. 5⅜ × 8½.
65084-7 Pa. $6.95

LAMINAR BOUNDARY LAYERS, edited by L. Rosenhead. Engineering classic covers steady boundary layers in two- and three-dimensional flow, unsteady boundary layers, stability, observational techniques, much more. 708pp. 5⅜ × 8½.
65646-2 Pa. $15.95

LECTURES ON CLASSICAL DIFFERENTIAL GEOMETRY, Second Edition, Dirk J. Struik. Excellent brief introduction covers curves, theory of surfaces, fundamental equations, geometry on a surface, conformal mapping, other topics. Problems. 240pp. 5⅜ × 8½.
65609-8 Pa. $6.95

GEOMETRY OF COMPLEX NUMBERS, Hans Schwerdtfeger. Illuminating, widely praised book on analytic geometry of circles, the Moebius transformation, and two-dimensional non-Euclidean geometries. 200pp. 5⅜ × 8¼.
63830-8 Pa. $6.95

MECHANICS, J.P. Den Hartog. A classic introductory text or refresher. Hundreds of applications and design problems illuminate fundamentals of trusses, loaded beams and cables, etc. 334 answered problems. 462pp. 5⅜ × 8½. 60754-2 Pa. $8.95

TOPOLOGY, John G. Hocking and Gail S. Young. Superb one-year course in classical topology. Topological spaces and functions, point-set topology, much more. Examples and problems. Bibliography. Index. 384pp. 5⅜ × 8¼.
65676-4 Pa. $7.95

STRENGTH OF MATERIALS, J.P. Den Hartog. Full, clear treatment of basic material (tension, torsion, bending, etc.) plus advanced material on engineering methods, applications. 350 answered problems. 323pp. 5⅜ × 8½. 60755-0 Pa. $7.50

ELEMENTARY CONCEPTS OF TOPOLOGY, Paul Alexandroff. Elegant, intuitive approach to topology from set-theoretic topology to Betti groups; how concepts of topology are useful in math and physics. 25 figures. 57pp. 5⅜ × 8½.
60747-X Pa. $2.95

ADVANCED STRENGTH OF MATERIALS, J.P. Den Hartog. Superbly written advanced text covers torsion, rotating disks, membrane stresses in shells, much more. Many problems and answers. 388pp. 5⅜ × 8½. 65407-9 Pa. $8.95

COMPUTABILITY AND UNSOLVABILITY, Martin Davis. Classic graduate-level introduction to theory of computability, usually referred to as theory of recurrent functions. New preface and appendix. 288pp. 5⅜ × 8½. 61471-9 Pa. $6.95

GENERAL CHEMISTRY, Linus Pauling. Revised 3rd edition of classic first-year text by Nobel laureate. Atomic and molecular structure, quantum mechanics, statistical mechanics, thermodynamics correlated with descriptive chemistry. Problems. 992pp. 5⅜ × 8½. 65622-5 Pa. $18.95

AN INTRODUCTION TO MATRICES, SETS AND GROUPS FOR SCIENCE STUDENTS, G. Stephenson. Concise, readable text introduces sets, groups, and most importantly, matrices to undergraduate students of physics, chemistry, and engineering. Problems. 164pp. 5⅜ × 8½. 65077-4 Pa. $5.95

THE HISTORICAL BACKGROUND OF CHEMISTRY, Henry M. Leicester. Evolution of ideas, not individual biography. Concentrates on formulation of a coherent set of chemical laws. 260pp. 5⅜ × 8½. 61053-5 Pa. $6.00

THE PHILOSOPHY OF MATHEMATICS: An Introductory Essay, Stephan Körner. Surveys the views of Plato, Aristotle, Leibniz & Kant concerning propositions and theories of applied and pure mathematics. Introduction. Two appendices. Index. 198pp. 5⅜ × 8½. 25048-2 Pa. $5.95

THE DEVELOPMENT OF MODERN CHEMISTRY, Aaron J. Ihde. Authoritative history of chemistry from ancient Greek theory to 20th-century innovation. Covers major chemists and their discoveries. 209 illustrations. 14 tables. Bibliographies. Indices. Appendices. 851pp. 5⅜ × 8½. 64235-6 Pa. $15.95

THE FOUR-COLOR PROBLEM: Assaults and Conquest, Thomas L. Saaty and Paul G. Kainen. Engrossing, comprehensive account of the century-old combinatorial topological problem, its history and solution. Bibliographies. Index. 110 figures. 228pp. 5⅜ × 8½.   65092-8 Pa. $6.00

CATALYSIS IN CHEMISTRY AND ENZYMOLOGY, William P. Jencks. Exceptionally clear coverage of mechanisms for catalysis, forces in aqueous solution, carbonyl- and acyl-group reactions, practical kinetics, more. 864pp. 5⅜ × 8½.   65460-5 Pa. $18.95

PROBABILITY: An Introduction, Samuel Goldberg. Excellent basic text covers set theory, probability theory for finite sample spaces, binomial theorem, much more. 360 problems. Bibliographies. 322pp. 5⅜ × 8½.   65252-1 Pa. $7.95

LIGHTNING, Martin A. Uman. Revised, updated edition of classic work on the physics of lightning. Phenomena, terminology, measurement, photography, spectroscopy, thunder, more. Reviews recent research. Bibliography. Indices. 320pp. 5⅜ × 8¼.   64575-4 Pa. $7.95

PROBABILITY THEORY: A Concise Course, Y.A. Rozanov. Highly readable, self-contained introduction covers combination of events, dependent events, Bernoulli trials, etc. Translation by Richard Silverman. 148pp. 5⅜ × 8¼.
63544-9 Pa. $4.50

THE CEASELESS WIND: An Introduction to the Theory of Atmospheric Motion, John A. Dutton. Acclaimed text integrates disciplines of mathematics and physics for full understanding of dynamics of atmospheric motion. Over 400 problems. Index. 97 illustrations. 640pp. 6 × 9.   65096-0 Pa. $16.95

STATISTICS MANUAL, Edwin L. Crow, et al. Comprehensive, practical collection of classical and modern methods prepared by U.S. Naval Ordnance Test Station. Stress on use. Basics of statistics assumed. 288pp. 5⅜ × 8½.
60599-X Pa. $6.00

WIND WAVES: Their Generation and Propagation on the Ocean Surface, Blair Kinsman. Classic of oceanography offers detailed discussion of stochastic processes and power spectral analysis that revolutionized ocean wave theory. Rigorous, lucid. 676pp. 5⅜ × 8½.   64652-1 Pa. $14.95

STATISTICAL METHOD FROM THE VIEWPOINT OF QUALITY CONTROL, Walter A. Shewhart. Important text explains regulation of variables, uses of statistical control to achieve quality control in industry, agriculture, other areas. 192pp. 5⅜ × 8½.   65232-7 Pa. $6.00

THE INTERPRETATION OF GEOLOGICAL PHASE DIAGRAMS, Ernest G. Ehlers. Clear, concise text emphasizes diagrams of systems under fluid or containing pressure; also coverage of complex binary systems, hydrothermal melting, more. 288pp. 6½ × 9¼.   65389-7 Pa. $8.95

STATISTICAL ADJUSTMENT OF DATA, W. Edwards Deming. Introduction to basic concepts of statistics, curve fitting, least squares solution, conditions without parameter, conditions containing parameters. 26 exercises worked out. 271pp. 5⅜ × 8½.   64685-8 Pa. $7.95

DE RE METALLICA, Georgius Agricola. The famous Hoover translation of greatest treatise on technological chemistry, engineering, geology, mining of early modern times (1556). All 289 original woodcuts. 638pp. 6¾ × 11.
60006-8 Clothbd. $15.95

SOME THEORY OF SAMPLING, William Edwards Deming. Analysis of the problems, theory and design of sampling techniques for social scientists, industrial managers and others who find statistics increasingly important in their work. 61 tables. 90 figures. xvii + 602pp. 5⅜ × 8½.
64684-X Pa. $14.95

THE VARIOUS AND INGENIOUS MACHINES OF AGOSTINO RAMELLI: A Classic Sixteenth-Century Illustrated Treatise on Technology, Agostino Ramelli. One of the most widely known and copied works on machinery in the 16th century. 194 detailed plates of water pumps, grain mills, cranes, more. 608pp. 9 × 12.
25497-6 Clothbd. $34.95

LINEAR PROGRAMMING AND ECONOMIC ANALYSIS, Robert Dorfman, Paul A. Samuelson and Robert M. Solow. First comprehensive treatment of linear programming in standard economic analysis. Game theory, modern welfare economics, Leontief input-output, more. 525pp. 5⅜ × 8½.
65491-5 Pa. $12.95

ELEMENTARY DECISION THEORY, Herman Chernoff and Lincoln E. Moses. Clear introduction to statistics and statistical theory covers data processing, probability and random variables, testing hypotheses, much more. Exercises. 364pp. 5⅜ × 8½.
65218-1 Pa. $8.95

THE COMPLEAT STRATEGYST: Being a Primer on the Theory of Games of Strategy, J.D. Williams. Highly entertaining classic describes, with many illustrated examples, how to select best strategies in conflict situations. Prefaces. Appendices. 268pp. 5⅜ × 8½.
25101-2 Pa. $5.95

MATHEMATICAL METHODS OF OPERATIONS RESEARCH, Thomas L. Saaty. Classic graduate-level text covers historical background, classical methods of forming models, optimization, game theory, probability, queueing theory, much more. Exercises. Bibliography. 448pp. 5⅜ × 8¼.
65703-5 Pa. $12.95

CONSTRUCTIONS AND COMBINATORIAL PROBLEMS IN DESIGN OF EXPERIMENTS, Damaraju Raghavarao. In-depth reference work examines orthogonal Latin squares, incomplete block designs, tactical configuration, partial geometry, much more. Abundant explanations, examples. 416pp. 5⅜ × 8¼.
65685-3 Pa. $10.95

THE ABSOLUTE DIFFERENTIAL CALCULUS (CALCULUS OF TENSORS), Tullio Levi-Civita. Great 20th-century mathematician's classic work on material necessary for mathematical grasp of theory of relativity. 452pp. 5⅜ × 8½.
63401-9 Pa. $9.95

VECTOR AND TENSOR ANALYSIS WITH APPLICATIONS, A.I. Borisenko and I.E. Tarapov. Concise introduction. Worked-out problems, solutions, exercises. 257pp. 5⅜ × 8¼.
63833-2 Pa. $6.95

CHALLENGING MATHEMATICAL PROBLEMS WITH ELEMENTARY SOLUTIONS, A.M. Yaglom and I.M. Yaglom. Over 170 challenging problems on probability theory, combinatorial analysis, points and lines, topology, convex polygons, many other topics. Solutions. Total of 445pp. 5⅜ × 8½. Two-vol. set.

Vol. I 65536-9 Pa. $5.95
Vol. II 65537-7 Pa. $5.95

FIFTY CHALLENGING PROBLEMS IN PROBABILITY WITH SOLUTIONS, Frederick Mosteller. Remarkable puzzlers, graded in difficulty, illustrate elementary and advanced aspects of probability. Detailed solutions. 88pp. 5⅜ × 8½.
65355-2 Pa. $3.95

EXPERIMENTS IN TOPOLOGY, Stephen Barr. Classic, lively explanation of one of the byways of mathematics. Klein bottles, Moebius strips, projective planes, map coloring, problem of the Koenigsberg bridges, much more, described with clarity and wit. 43 figures. 210pp. 5⅜ × 8½.
25933-1 Pa. $4.95

RELATIVITY IN ILLUSTRATIONS, Jacob T. Schwartz. Clear non-technical treatment makes relativity more accessible than ever before. Over 60 drawings illustrate concepts more clearly than text alone. Only high school geometry needed. Bibliography. 128pp. 6⅛ × 9¼.
25965-X Pa. $5.95

AN INTRODUCTION TO ORDINARY DIFFERENTIAL EQUATIONS, Earl A. Coddington. A thorough and systematic first course in elementary differential equations for undergraduates in mathematics and science, with many exercises and problems (with answers). Index. 304pp. 5⅜ × 8¼.
65942-9 Pa. $7.95

FOURIER SERIES AND ORTHOGONAL FUNCTIONS, Harry F. Davis. An incisive text combining theory and practical example to introduce Fourier series, orthogonal functions and applications of the Fourier method to boundary-value problems. 570 exercises. Answers and notes. 416pp. 5⅜ × 8½.
65973-9 Pa. $8.95

THE THOERY OF BRANCHING PROCESSES, Theodore E. Harris. First systematic, comprehensive treatment of branching (i.e. multiplicative) processes and their applications. Galton-Watson model, Markov branching processes, electron-photon cascade, many other topics. Rigorous proofs. Bibliography. 240pp. 5⅜ × 8½.
65952-6 Pa. $6.95

AN INTRODUCTION TO ALGEBRAIC STRUCTURES, Joseph Landin. Superb self-contained text covers "abstract algebra": sets and numbers, theory of groups, theory of rings, much more. Numerous well-chosen examples, exercises. 247pp. 5⅜ × 8½.
65940-2 Pa. $6.95

GAMES AND DECISIONS: Introduction and Critical Survey, R. Duncan Luce and Howard Raiffa. Superb non-technical introduction to game theory, primarily applied to social sciences. Utility theory, zero-sum games, n-person games, decision-making, much more. Bibliography. 509pp. 5⅜ × 8½. 65943-7 Pa. $10.95
